*POLYMERS
FOR GAS
SEPARATION*

POLYMERS FOR GAS SEPARATION

Naoki Toshima, Editor
Department of Industrial Chemistry
Faculty of Engineering
The University of Tokyo

Naoki Toshima
Department of Industrial Chemistry
Faculty of Engineering
University of Tokyo
Hongo, Bunkyo-ku
Tokyo, Japan

Library of Congress Cataloging-in-Publication Data

Polymers for gas separation / edited by Naoki Toshima.
   p. cm.
   Includes bibliographical references and index.
   ISBN 1-56081-093-9
   1. Gases—Separation. 2. Gas separation membranes.
3. Polymers. I. Toshima, Naoki.
TP242.P65 1992
660'.2842—dc20   92-2645
                                                                    CIP

©1992 VCH Publishers, Inc.
This work is subject to copyright.
   All rights are reserved, whether the whole or part of the material is concerned, specifically those of translation, reprinting, re-use of illustrations, broadcasting, reproduction by photocopying machine or similar means, and storage in data banks.
   Registered names, trademarks, etc., used in this book, even when not specifically marked as such, are not to be considered unprotected by law.

Printed in the United States of America
ISBN 1-56081-093-9 VCH Publishers
ISBN 3-527-28234-3 VCH Verlagsgesellschaft

Printing History:
10 9 8 7 6 5 4 3 2 1

Published jointly by:

VCH Publishers, Inc.
220 East 23rd Street
Suite 909
New York, New York 10010

VCH Verlagsgesellschaft mbH
P.O. Box 10 11 61
D-6940 Weinheim
Federal Republic of Germany

VCH Publishers (UK) Ltd.
8 Wellington Court
Cambridge CB1 1HW
United Kingdom

Preface

Separation is an important process in chemistry. Let us consider a chemical reaction: Reactants must be purified by separation of impurities, and the product must be purified by separation of reactants and byproducts from the reaction mixtures. In fact, most chemical-synthetic processes involve separation. On the other hand, gases are simple molecules and are produced industrially in the largest quantities among all chemicals throughout the world. Therefore, gas separation is a most important process, not only in academic research, but also in industry.

Polymers play an important role in gas separation. Polymers form membranes that are used for gas separation by physical means, and they also form polymer–metal complexes, which serve as adsorbants for gaseous molecules and can separate gases by chemical means.

This book consists of three parts: the Introduction, Separation by Physical Methods, and Separation by Chemical Methods. In the introduction the requirements and general methods for gas separation are mentioned briefly. The characteristics of polymers and future prospects for gas separation using polymers are then reviewed.

Membrane separation is the technique most widely investigated and actually utilized industrially. The second part of this book is concerned with gas separation by this method. After describing the theory of membrane permeation, two new kinds of membranes are described from the viewpoint of the relationship of aggregation state to gas permeation properties and the design of polymer membranes. These chapters on new polymer membranes will provide fresh ideas for breakthroughs in gas separation by polymers.

Polymer complexes are described in Part III. The concept of gas separation by metal complexes is not new, but the use of polymer–metal complexes has been practically realized on the basis of general information on these complexes, which have recently been developed all over the world, especially in Japan. Polymer complexes can provide advanced new materials that are intelligent and multifunctional.

If this volume succeeds in serving the scientific community as introduction for scientists preparing to enter the field, as a reference book for researchers already in it and seeking new ideas, and as a textbook for students wanting to learn an advanced application of new polymer materials, then our efforts will not have been in vain.

Last, but not least, I wish to acknowledge the contributors for their efforts to provide excellent material, and the publisher and the editors, especially Dr. Charles H. Doering, Executive Editor, and Mr. Mark Sacher, Production Editor, for their patience, encouragement, and assistance with grammatical improvements.

July 30, 1991   Naoki Toshima

Contents

Contributors   ix

Part I. Introduction

1 Introduction to Gas Separation by Polymers   3
   Naoki Toshima

Part II. Separation by Physical Method

2 Theory of Membrane Permeation   15
   Shoji Kimura and Takuji Hirose

3 Relationships between Aggregation State and Gas Permeation Properties   51
   Tisato Kajiyama

4 Design of Polymer Membranes for Gas Separation   107
   Hisashi Odani and Toshio Masuda

Part III. Separation by Chemical Methods

5 Porous Polymer Complexes for Gas Separation   147
   Naoki Toshima and Hiroyuki Asanuma

6   Polymer Complex Membranes for Gas Separation   **183**
Hiroyuki Nishide and Eishun Tsuchida

7   Polymer Complex for the Separation of Carbon Monoxide and Ethylene   **221**
Hidefumi Hirai

Index   **239**

Contributors

Hiroyuki Asanuma
Ashigara Research Laboratories
Fuji Photo Film Co. Ltd.
Nakanuma, Minamiashigara
Kanagawa 250-01, Japan

Hidefumi Hirai
Department of Industrial
Chemistry
Faculty of Engineering
Science University of Tokyo
Kagurazaka, Shinjuku-ku
Tokyo 162, Japan

Takuji Hirose
Industrial Products Research
Institute
Agency of Industrial Science and
Technology
MITI
Higashi, Tsukuba
Ibaraki 305, Japan

Tisato Kajiyama
Department of Chemical Science
and Technology
Faculty of Engineering
Kyushu University
Hakozaki, Higashi-ku
Fukuoka 812, Japan

Shoji Kimura
Department of Chemical
Engineering
Faculty of Engineering
The University of Tokyo
Hongo, Bunkyo-ku
Tokyo 113, Japan

Toshio Masuda
Department of Polymer Chemistry
Faculty of Engineering
Kyoto University
Yoshida Honmachi
Sakyo-ku
Kyoto 606-01, Japan

Hiroyuki Nishide
Department of Polymer Chemistry
Waseda University
Okubo, Shinjuku-ku
Tokyo 169, Japan

Hisashi Odani
Institute for Chemical Research
Kyoto University
Uji
Kyoto 611, Japan

Naoki Toshima
Department of Industrial
Chemistry
Faculty of Engineering
The University of Tokyo
Hongo, Bunkyo-ku
Tokyo 113, Japan

Eishun Tsuchida
Department of Polymer Chemistry
Waseda University
Okubo, Shinjuku-ku
Tokyo 169, Japan

Part I

Introduction

1 Introduction to Gas Separation by Polymers

Naoki Toshima

1. Need for Gas Separation
2. Gas-Separation Methods
   2.1. The Cryogenic Method
   2.2. The Membrane Method
   2.3. The Sorption Method
3. Gas Separation and Polymers—Characteristics and Future Prospects

1. Need for Gas Separation

Isolated gases have been used for many years for various purposes, as one may experience in daily life. For example, oxygen may be used for inhalation by hospital patients or in welding by an iron worker. Rubber balloons as well as large advertising balloons and airships are filled with a light gas, helium. Carbonated drinks and beer contain carbon dioxide. Solid carbon dioxide, called dry ice, is used as a refrigerant. Nitrogen is used as a space-filler in juice and food containers because it is an inert gas and can keep the contents fresh. Polyfluorinated hydrocarbons are used as the coolant in refrigerators.

Table 1.1 shows the top ten chemical products in the United States[1] and in Japan.[2] Among the top ten chemicals in the United States, six (i.e., nitrogen, oxygen, ethylene, ammonia, propylene, and chlorine) are gaseous under usual conditions. A similar situation is observed in Japan. Attention

Table 1.1 Top Ten Chemical Products in the United States and Japan

Chemical	United States[a]		Japan[b]	
	Rank	Production (tt y^{-1})	Rank	Production (tt y^{-1})
Sulfuric acid	1	44,281	4	6,885
Nitrogen	2	28,660	3	9,196
Oxygen	3	19,495	1	11,179
Ethylene	4	18,740	5	5,603
Lime	5	17,400	2	10,456
Ammonia	6	16,958	15	1,831
Phosphoric acid	7	12,175	—	169
Sodium hydroxide	8	11,688	9	3,674
Propylene	9	11,060	8	4,036
Chlorine	10	10,942	7	4,320
Calcium sulfate	—	—	6	4,548
Benzene	16	5,930	10	2,903
Carbon dioxide	17	5,490	19	1,344

[a] In 1991 according to Ref. 1. tt = thousand ton.
[b] In 1990 according to Ref. 2.

should be paid to the fact that the fundamental mass-produced chemicals are mostly gaseous molecules. Thus it can be expected that the market for gas separation will continue to grow.

The gas separation performed on the largest scale is that of nitrogen and oxygen from air, which is mainly performed cryogenically. However, separation by a polymer membrane is interesting, especially from the viewpoint of energy economy.[3] Monsanto claims that its Prism units were the first large-scale gas separation membrane systems, introduced in 1977. Since then there have been many applications of polymer membranes to gas separation processes, the principal including the recovery of hydrogen from refinery streams for recycling, the removal of carbon dioxide and hydrogen sulfide from natural gas, the production of nitrogen and enriched air, the recovery of ethylene or propylene from chemical plants for recycling, the recovery of methane from mines and landfills, and the recovery of certain gases from exhausts. Future large-scale systems will be devoted to health care, food processing, and biotechnology.

Although the usual separation processes deal with the gases at rather high concentrations, there is another type that is concerned with gases at low concentrations. For example, the exhaust gas from power plants and generation stations usually contains of the order of 1000 ppm sulfur dioxide and nitrogen oxide. The volume of exhaust gas needing treatment is enormous, but the amount of the gas that should be removed from the exhaust is very small. Nevertheless, the complete removal of the gas is required by environmental regulations, since sulfur dioxide and the nitrogen oxides are responsible for acid rain, in addition to contributing to ozone pollution. In practice sulfur dioxide and nitrogen oxides are removed with the combination of individual processes (i.e., by wet absorption and by selective catalytic reduction) (SCR, dry process) using ammonia as a reductant, respectively. Therefore, it is desirable to develop new methods that can be applied to this kind of gas separation (e.g., the simultaneous removal of sulfur dioxide and nitrogen oxides from the exhaust gas).

The separation of carbon dioxide from exhaust gas will soon be required from the viewpoint of the global environment. Although the concentration of carbon dioxide is larger than that of sulfur dioxide or nitrogen oxide, the removal of carbon dioxide belongs in the same category as that of sulfur dioxide or the nitrogen oxides in large scale. In this case the separation must be performed with less energy consumption, since the consumption of energy produces extra carbon dioxide.

The removal of a small amount of gas is also very important for industry. Although the individual quantities are not large, the removal of oxygen from food packaging can keep the food fresh by preventing the oxidation of the food. The removal of a trace amount of ethylene from packaged fruit retards ripening, which makes the long-distance transportation and the long-term storage possible.

Examples of gas-separation applications can be classified as shown in Table 1.2.

2. Gas-Separation Methods

Separation is a very important process in chemistry. Chemicals can usually be separated by differences in their physical properties, and sometimes by their chemical properties. In other words, separation can be performed mostly by physical methods and occasionally by chemical methods. This is also true for the separation of gaseous molecules.

2.1. The Cryogenic Method

Most practical gas separations can be performed by the boiling-point differences in the gaseous molecules. Since the boiling point of gaseous molecules is below room temperature, the separation must be carried out at low temperatures, and is called a *cryogenic method*. In this process gases are separated through fractional condensation and distillation.

The cryogenic method was first applied to the separation of air into oxygen and nitrogen by Linde in 1910. The gases obtained at relatively low cost by the cryogenic separation of air are now used extensively in the electronics industry as high-purity inert gases (N_2, Ar), in the steel industry for BOF (Bessemer Oxygen Furnace) processes, and in municipal wastewater treatment to increase processing capacity (O_2). Liquid nitrogen is used as a refrigerant in the food industry for efficient freezing as well as for sustained refrigeration during transportation. Helium, hydrogen, and argon

Table 1.2 Categories of Gas Separation

Category	Scale	Gas	Examples
Purification or concentration	Large scale	Hydrogen	Recovery of H_2 from refinery systems and chemical plants
		Nitrogen	Production of N_2 from air, turbine exhaust gas, and boiler flue gas for food processing, food transport, metal-working surroundings, and blanketing of fuel tanks
		Oxygen	Production of O_2 and enriched air from air for health care, and effective combustion
		Methane	Recovery of CH_4 from mines and landfills
		Carbon monoxide	Recovery and purification of CO from steel gas and water gas
		Carbon dioxide	Separation and recovery of CO_2 from power plants, fuel gas, and recovery gas in tertiary oil production
		Oxygen	Purification of nitrogen by removel of O_2
Removal	Large scale	Nitrogen monoxide Sulfur dioxide	Treatment of exhaust gas from power plants
		Carbon dioxide	Removal of CO_2 from natural gas and exhaust gas
		Hydrogen sulfide	Removal of H_2S from natural gas
		Hydrogen chloride	Removal of HCl from flue gas and exhaust gas
		Furon	Removal of furon from refrigerators
	Small scale	Oxygen	Removal of O_2 from packaging, tins, and bottles for food storage
		Ethylene	Removal of C_2H_4 from fruit packaging

Gas-Separation Methods

can also be separated by cryogenic means. The technology using these gases has contributed greatly to scientific research and has also achieved widespread industrial use.

Although the method has been extensively used in industry, it does have disadvantages. In air separation, for example, the water and carbon dioxide included in the air must be removed by prefreezing or by trapping with adsorbants, since they deposit upon cooling and block pipes.

2.2. The Membrane Method

Membrane separation is a noncryogenic method that is increasingly employed even for large-scale systems.[3] In this process the difference in solubility and diffusion of the gaseous molecules in the polymer membrane is the driving force for the separation, as illustrated in Fig. 1.1. The separation system for hydrogen was first developed by du Pont

Figure 1.1 Model of gas separation with a polymer membrane.

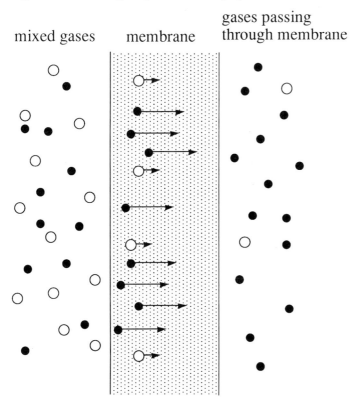

○ gaseous molecule with low solubility into membrane
● gaseous molecule with high solubility into membrane

as "Permasep" in 1970 and then by Monsanto as "Prism" in 1977. In 1985 "Generon" was developed by Dow Chemicals for the separation of nitrogen.

Although the gas selectivity inherent in membrane separation is not high, the low energy cost of the method is very attractive. Because of this advantage, Japan has become interested in the technique and has started research on polymer membranes for oxygen enrichment as a national project in the 1980s after the oil crisis.

The materials used in the membranes are synthetic polymers, examples of which are listed in Table 1.3.[4] The polymers should be mechanically strong, and have high permeability and selectivity. The details of the theory and applications of polymer membranes will be described in Part II in this book. Gas permeability and selectivity in polymeric membranes are usually inversely proportional. Researchers are making an effort to increase permeability while maintaining selectivity, but it is difficult to develop new polymer membranes having such properties. Asymmetric membranes and composite membranes have been developed to solve this problem. Using polymer–metal complexes may provide an answer to this problem, and details of such membranes will be described in Chapter 6.

Membrane separation requires a large surface area, which can be provided by hollow fibers or spiral-wound sheets. Recent developments in membrane technology involve not only new polymeric materials but also new systems. However, a description of the latter is outside the scope of this book.

2.3. The Sorption Method

Sorption phenomena can also be used for the separation of gases. The adsorbed gas can be desorbed by increasing temperature or decreasing pressure. For example, the differential adsorption of hydrogen or nitrogen on molecular sieves is used for rather small-scale separation processes, which is known as a pressure swing adsorption (PSA) method. In 1966 UCC developed the hydrogen-PSA, and in 1973 Bergbau-Forschung developed the oxygen-PSA. A PSA plant for oxygen, which can be started quickly and easily shut down, can be used for small hospitals as well as units producing more than 25,000 m^3/d.

Chemical sorption can also be used for selective separation of gases. Some kinds of metal complexes have coordination properties that can "catch" a special kind of gaseous molecule. These properties can be used in gas separation. For example, oxygen coordination to iron(II) or cobalt(II) complexes can be applied to the removal and selective transportation of oxygen by the immobilization of the complexes into polymers.

Such immobilization has been investigated for the purpose of developing synthetic enzymes or heterogenized metal-complex catalysts.[5–7] Cumulative techniques as well as knowledge of the polymer–metal complex

Table 1.3 Permeability Coefficients of Various Polymer Membranes[a]

Polymer membrane	Temp. (°C)	Permeability coefficient[b]				
		He	H_2	CO_2	O_2	N_2
Polydimethylsiloxane	25	230		3240	605	300
Poly(4-methyl-1-pentene)	25	100		93	32	
Natural rubber	25	23.7	90.8	99.6	17.7	6.12
Ethyl cellulose	25	53.4		113	15	4.43
Poly(2,6-dimethylphenylene oxide)	25			75	15	3.0
Polytetrafuroroethylene	25			12.7	4.9	
Polyethylene (low density $d = 0.922$)	25	4.93		12.6	2.89	0.97
Polystyrene	20	16.7		10.0	2.01	0.32
Polycarbonate	25	19		8.0	1.4	0.30
Butyl rubber	25	8.42		5.2	1.30	0.33
Cellulose acetate	22	13.6			0.43	0.14
Polypropylene (2-dimensionally oriented)	27			1.8	0.77	0.18
Polyethylene (high density $d = 0.964$)	25	1.14		3.62	0.41	0.143
Poly(vinyl chloride) (30% DOP)	25	14.0	13	3.7	0.60	0.20
Nylon-6	30	0.53		0.16	0.038	
Poly(ethylene telephthalate)	25	1.1	0.6	0.15	0.03	0.006
Poly(vinylidene chloride)	25		0.08	0.029	0.005	0.001
Polyacrylonitrile	25	0.55		0.0018	0.0003	
Poly(vinyl alcohol)	20	0.0033		0.0005	0.00052	0.00045

[a] Ref. 4.
[b] cm^3 (STP) (cm/cm^2) sec (cm Hg)

are now being applied to gas separation, the details of which will be described in Part III.

Removal of a trace amount of gases is an important technique for industries. For example, the mixtures of powders of metallic iron are extensively used for removal of oxygen from food packagings. However, this technique has disadvantage for food, since the process usually involves a magnetic checking system to exclude undesired steel-fragments. The metal complex adsorbants have a potential to be used to remove a trace amount of oxygen from the food packagings.

3. Gas Separation and Polymers—Characteristics and Future Prospects

Among the gas-separation methods mentioned, a cryogenic method has been applied to simple and large-scale separations. However, rather complicated separations need membrane and/or sorption methods, in which organic polymers play important roles.

Polymers have several advantages as materials:

1. Organic polymers are light in weight and processible to any form, such as thin films and porous beads.
2. Polymers have huge variations in structure and properties. Many kinds of monomers having various functional groups can be used for polymerization. Plural numbers of monomers can form copolymers in block, random, and regulated ways. Graft copolymerization gives different copolymers from the same monomers. The introduction of functional groups by polymer reactions is an important technique in preparing new polymers.
3. Polymers can form composite materials. Various combinations are available by using organic polymers, asymmetric polymer membranes and polymer–metal composite membranes being examples.

Because of these advantages of organic polymers, polymeric membranes are now increasingly being used for the separation not only of gaseous but also of liquid molecules on the large scale. Attention must be paid to the fact that membrane separation can be carried out with a low energy cost. For this purpose, various kinds of polymeric membranes have been developed, and now this technique is arriving at a new stage.

From the viewpoint of the global environment, the need for carbon dioxide separation is increasing rapidly. Now the investigation of new polymeric membrane materials is being carried out using fluorinated polymers with high gas solubility, aromatic polyamides with high gas permeability, asymmetric cellulose acetates with high flexibility, siloxane copolymers with high gas permeability, basic polymers like polyamine with high CO_2 interaction, and polysulfones with high CO_2 solubility.

The technology needs a breakthrough in the membrane materials not only regarding permeability, but also selectivity, as well as durability.

As mentioned, the organic polymers are easily designed, which makes it possible to form polymer–metal complexes. Thus the functional groups involved in polymers can serve as ligands to form a strong coordination bond between the polymer and the metal ion, producing the polymer–metal complex.

By forming this complex, the active metals can be easily handled and attain the advantages of polymeric materials. Moreover, the polymer can affect the metal complex as follows[7]:

1. Diluting effect: Polymers can immobilize each metal complex separately, and then inhibit the easy movement of the metal complex, which prevents collisions of the metal complexes with each other.

2. Concentrating effect: Immobilizing metal complexes into polymers can make the concentration of the metal complexes in the polymer higher than in the solution, which makes cooperation possible.

3. Field effect: Polymers can form microdomains, which may provide the reaction field for the reaction occurring around the metal complex. These polymer fields can promote or inhibit the approach of individual molecules selectively.

4. Steric effect: The polymers surrounding the metal complex can affect the approach of the individual molecules sterically.

As the result of these effects, the polymer–metal complexes are more stable, more effective, and more easily handled than solutions of the corresponding metal complexes for gas separation. These relatively new materials are expected to be used as both membranes and adsorbants. Since the coordination of gaseous molecules to the metal complex is stronger than the physical adsorption or solubilization into polymers, the polymer–metal complex provides higher selectivity for gas separation. Therefore, polymer–metal complexes will be useful for complete separation or removal of a trace amount of a particular gas. High permeability, when used in membrane separation, and high desorbing ability, when used in sorption, will be a key point for the development of the corresponding polymeric materials.

REFERENCES

1. *Chem. Eng. News,* April 8, 1991, p. 14.
2. *Year Book of Chemical Industries Statistics, 1989.* Research and Statistics Dept., MITI.
3. J. Haggin, *Chem. Eng. News,* June 6, 1988, pp. 7–16.
4. T. Nakagawa, *Makubunri Gijutsu Taikei* (*Encyclopedia of Membrane Separation Technology*), Vol. 1, Fuji Technosystem, Tokyo, 1991, p. 28.

5. N. Toshima, M. Kaneko, and M. Sekine, *Kobunshi Sakutai* (*Polymer Complexes*), Kyoritsu, Tokyo, 1990.
6. E. Tsuchida, Ed., *Macromolecular Complexes—Dynamic Interactions and Electronic Processes*, VCH, New York, 1991.
7. H. Hirai and N. Toshima, in *Tailored Metal Catalysts*, Y. Iwasawa, Ed., Reidel, Dordrecht, 1986, pp. 87–140.

Part II

Separation by Physical Method

ptim
2. Theory of Membrane Permeation

Shoji Kimura and Takuji Hirose

1. Transport Phenomena of Gas Permeation through Membranes
   1.1. Gas Transport through a Porous Membrane
   1.2. Gas Transport through a Nonporous Membrane
   1.3. Gas Solution in Polymers
   1.4. Gas Diffusion in Polymers
2. Process Design of Gas-Separation Schemes
   2.1. Separation Factor
   2.2. Separation Stage
   2.3. Separation Cascade
   2.4. New Scheme (1)
   2.5. New Scheme (2)

1. Transport Phenomena of Gas Permeation through Membranes

1.1. Gas Transport through a Porous Membrane

The permeation of a gas through a porous polymer membrane is generally described by transport equations through capillary tubes based on the kinetic theory of gases.

Three types of mechanisms have been proposed for gas transport through capillaries, namely, *Poiseuille* (or viscous) *flow*, *Knudsen* (or free-molecule) *flow*, and *slip flow*. Depending on the relative magnitude of pore radius of a capillary r and mean free path λ of the gas, gas molecules pass through the capillary at the given pressure and temperature by any one of the mechanisms stated above. The mean free path is obtained by the relation:

$$\lambda = \frac{RT}{2^{1/2} \pi d^2 N(p_1 - p_2)}, \tag{1}$$

where R is the universal gas constant, T is the absolute temperature, d is the collision diameter of the penetrant gas, N is the Avogadro number, and p_1 and p_2 are the upstream and downstream penetrant pressures, respectively, across the membrane. For a single capillary, the following equations have been derived for the three mechanisms of flow.[1]

When $r \gg \lambda$, the mechanism is Poiseuille flow, and the flow rate or flux G is described by the form:

$$G = \frac{\pi r^4}{8 \eta RT \delta} \frac{p_1 + p_2}{2} (p_1 - p_2), \tag{2}$$

where G is the quantity of transported gas in the gas phase, η is the coefficient of viscosity of the penetrant gas, and δ is the length of a capillary tube.

When $r \ll \lambda$, on the other hand, Knudsen flow described by the following form is predominant:

$$G = \frac{8\pi r^3 (p_1 - p_2)}{3 \delta (2\pi MRT)^{1/2}}, \tag{3}$$

where M is the molecular weight of the penetrant gas.

In the intermediate range, where r and λ are comparable, slip flow is introduced as a correction term, and the equation is:

$$G = \frac{\pi r^3}{M\bar{v} \delta} \frac{(p_1 - p_2)}{2}, \tag{4}$$

where $\bar{v}$, the mean speed, is obtained by the relation[1]

$$\bar{v} = \left(\frac{8RT}{\pi M} \right)^{1/2}. \tag{5}$$

The above equations (2)–(5) are applicable for a pure penetrant gas, and more effects on gas flows are considered for separation of a gas mixture. One of them is the Present–deBethune[2] effect, which is caused by the fact that the penetrant molecules have different masses and mean speeds. Another is caused by the opening of capillaries when the radius is of the order of the gas molecules' radii. These capillaries sieve penetrant gases by their sizes, and the flow is called *restricted diffusion*. Although gas separation by this mechanism is effective, it has never been applied to real membranes.

The gas flows mentioned in the previous paragraphs are considered to be uninfluenced by the forces of interaction between the penetrant gas and the membrane material. It is also necessary to consider the flow resulting from the mobility of the layer of the adsorbed gas on the wall of a capillary, namely, "surface" flow. Gilliland et al. derived the following equation for surface flow:[3]

$$G = \frac{RT\,\rho_{app}A_p}{\tau^2 C_R S_s L_p} \int_{p_2}^{p_1} \frac{x^2}{p}\,dx, \tag{6}$$

where ρ_{app} is the apparent density of the membrane, A_p is the cross section of a capillary, τ is the tortuosity factor, C_R is the coefficient of resistance, S_s is the specific surface area of the solid over which the adsorbed molecules are mobile, L_p is the pore length, and x is the amount adsorbed per unit weight of the membrane.

1.2. Gas Transport through a Nonporous Membrane

The permeation of a gas through a nonporous polymer membrane is generally described in terms of a "solution-diffusion" mechanism.[4,5] Application of gas pressure on one side of a membrane leads to the following sequence of steps:[4]

1. Adsorption and solution of the gas at the interface of the membrane, a sorption process;
2. Random movement of the dissolved gas in and through the membrane, a diffusion process;
3. Release of the gas at the opposite interface, a desorption process.

The term *permeation* is used to describe the overall mass transport of the penetrant gas across the membrane, whereas the term *diffusion* refers only to the movement of the gas molecule inside the polymer membrane, step 2. The sorption and desorption steps, 1 and 3, are fast, and gas solution equilibrium is established at the membrane interfaces when constant gas pressures are maintained. In contrast, the diffusion step, 2, is very slow, and hence is the rate-determining step in the permeation process. All polymer membranes are observed to be permeable to various extents to all gases.

Diffusion of a penetrant gas in a membrane can be described by Fick's first law:[6,7]

$$G = -D(C)\,\nabla C, \tag{7}$$

where G is the rate of diffusion of the penetrant gas through a unit area of the membrane, D is the local diffusion coefficient for the present gas–membrane system and temperature and can either be a constant or a function of the concentration, and C is the local concentration of the penetrant gas. This equation takes the following form for one-dimensional transport in a direction x normal to an isotropic and homogeneous membrane:

$$G = -D(C)\frac{\partial C}{\partial x}. \tag{8}$$

The concentration gradient in the membrane can be obtained from Fick's second law, which takes the following form for one-dimensional diffusion:

$$\frac{\partial C}{\partial t} = \frac{\partial}{\partial t}\left(\frac{\partial C}{\partial x}\right), \tag{9}$$

where $(\partial C/\partial t)$ is the rate of change in concentration with time t at a position coordinate x. Solutions of this differential equation and comparable equations for a variety of membrane geometries have been summarized under various conditions by several researchers.[5,7,8]

The steady-state permeation flux can also be expressed in the following form in pressure units:[5,6,8]

$$G_s = \frac{\bar{P}(p_1 - p_2)}{l}, \tag{10}$$

where $\bar{P}$ is the mean permeability coefficient, p_1 and p_2 are the upstream and downstream penetrant pressures, respectively, across the membrane of thickness l.

At constant temperature, the concentration of a penetrant gas in a polymer membrane is determined by the solubility and the pressure. The relation between C and the pressure p at solution equilibrium is usually expressed in the form:

$$C = S(C)p, \tag{11}$$

where $S(C)$ is the solubility coefficient and is generally a function of concentration.

The steady-state flux through a planar, isotropic, and homogeneous membrane at constant temperature is obtained from Eqs. (8), (10), and (11) as:

$$\bar{P}(C) = \bar{D}(C)\bar{S}(C), \tag{12}$$

where $\bar{D}(C)$ and $\bar{S}(C)$ are the mean diffusion and solubility coefficients, respectively, and are defined as follows:

$$\bar{D}(C) = \frac{\left(\int_{C_2}^{C_1} D(C)dC\right)}{(C_1 - C_2)} \tag{13}$$

$$\bar{S}(C) = \frac{C_1 - C_2}{p_1 - p_2}. \tag{14}$$

The penetrant pressures p_1 and p_2 at the two membrane interfaces, respectively, produce the equilibrium concentrations C_1 and C_2 at each interface. The rate of permeation through a nonporous polymer membrane is usually reported in the form of P, which depends on the nature of the polymer and the penetrant gas, and generally on the penetrant pressure (concentration) and temperature. Some typical pressure-dependence forms of permeability coefficients are schematically presented in Fig. 2.1.[4,5] As seen from Eqs. (12)–(14), the pressure (concentration) dependence of P is determined by the pressure (concentration) dependences of D and S. To understand and discuss the gas transport or its mechanisms in a polymer membrane, therefore, it is necessary to investigate the diffusion and solu-

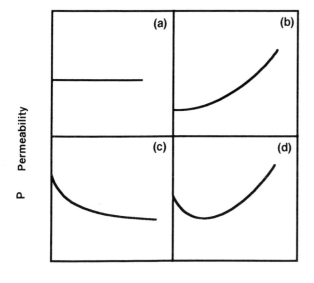

Figure 2.1 Schematic representation of some concentration-dependent forms of permeability coefficients. Typical for (a) gas-rubbery polymer systems, (b) vapor-rubbery polymer systems, (c) gas-glassy polymer systems, and (d) vapor-glassy polymer systems.

bility coefficients as well as the permeability coefficient, and their pressure (concentration) dependences.

It has been observed that both sorption and diffusion mechanisms strongly depend on the relationship between the temperature T and the glass transition temperature T_g of the polymer.[5,9] The schematic relationship between the temperature and the specific volume of a polymer is shown in Fig. 2.2. All polymers undergo a phase transformation from a soft and "rubbery" state to a hard or "glassy" state as the temperature is lowered below T_g.

At temperatures above T_g, the segmental motions of the rubbery polymer chains are rapid, so that they respond quickly to the environmental changes of the polymer. As a result, solution equilibrium between a rubbery polymer and a penetrant gas is established in a short time compared to the time required for the diffusion. Accordingly, all of the penetrant molecules are believed to behave in a unique mode. In contrast, at temperatures below T_g, the segmental motions of the glassy polymer chains are restricted and cannot completely homogenize the environment of a gas molecule. The glassy polymer itself is not in a true equilibrium state within the experimental time scale and involves the unrelaxed volume segments called "holes" or "microvoids" of different sizes. The inhomogeneity at a

Figure 2.2 Schematic representation of the relationship between temperature and specific volume of a polymer.

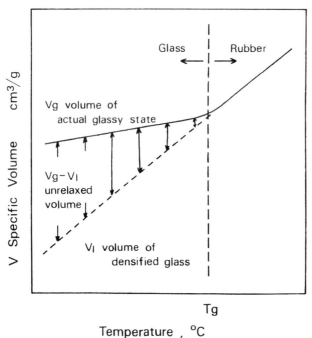

molecular scale is believed to cause different or multiple gas solution and diffusion modes.[5,9] In the following text, gas solution and diffusion behaviors in rubbery and glassy polymers are provided separately.

1.3. Gas Solution in Polymers

1.3.1. Rubbery Polymers

PRESSURE (CONCENTRATION) DEPENDENCE The concentration of a penetrant gas in a polymer is related to the pressure at solution equilibrium and constant temperature by Eq. (11).[6,7] The solubility coefficient depends on the nature of the polymer and the penetrant gas and is generally a function of penetrant concentration. Figure 2.3 schematically illustrates typical sorption (or solubility) isotherms, plots of concentration versus pressure for polymer–gas systems. A solution of gases in rubbery polymers yields sorption isotherms that are either linear [Fig. 2.3(a)] or convex to the pressure axis [Fig. 2.3(b)]. Linear sorption isotherms are obtained when the penetrant solution obeys the Henry's law, that is:

$$S(C) = S(0) \quad \text{(constant)}, \tag{15}$$

where $S(0)$ is the solubility coefficient at the limit of zero pressure. This is usually the case when the temperature is higher than the critical tempera-

Figure 2.3 Schematic representation of typical sorption isotherms. (a) Henry's law, (b) BET III (Flory-Huggins), (c) Dual-mode, (d) Dual-mode accompanied with glass transition, and (e) BET II (Dual-mode accompanied with glass transition).

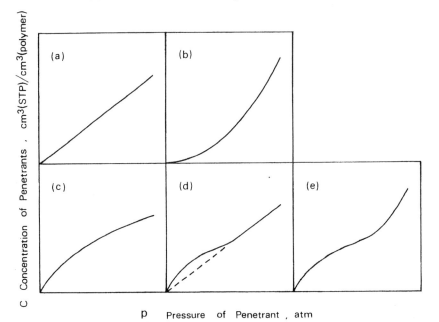

ture of the gas T_c $(T > T_c)$.[10–13] On the other hand, isotherms that are convex to the pressure axis are observed with gases that exhibit a high solubility in the polymer and plasticize the polymer at higher pressures (concentrations).[13–16] This is the case when T is close to or lower than T_c $(T < T_c)$, and the sorption isotherm is described by the Flory–Huggins equation:[17]

$$\ln(p/p_0) = \ln(v) + \ln(1 - v) + \chi(1 - v)^2, \tag{16}$$

where p is the pressure of the gas, p_0 is its vapor pressure at the temperature, v is the volume fraction of the dissolved gas, and χ is the Flory–Huggins parameter. This equation reduces to the exponential form when the penetrant concentration is low.[15]

$$C = [S(0) \exp(\alpha C)]p, \tag{17}$$

where α is a constant that characterizes the concentration dependency. Water vapor, which tends to cluster together in nonpolar polymers, sometimes exhibits a convex isotherm as in Fig. 2.3(b).[18–20]

TEMPERATURE DEPENDENCE   The temperature dependence of the solubility coefficient over small ranges of temperature can be represented by the van't Hoff–type relation[8,21–23]

$$S = S_0 \exp(-\Delta H_s/RT), \tag{18}$$

where S_0 is a constant and ΔH_s is the enthalpy of solution (the heat of solution). The solubility of penetrant gases in polymers commonly decreases with increasing temperature; that is, the solution process is exothermic; hence, ΔH_s is generally negative. However, ΔH_s also depends on the nature of a polymer, and the sign for smaller gases, such as H_2, He, and Ne, is often positive. The solubility of different gases in a rubbery polymer increases with its T_c, and T_c can be a scaling factor for the solubility.[24–26] When the solubility coefficient is pressure (concentration) dependent, $S(0)$ is usually used for S in Eq. (18).

1.3.2. Glassy Polymers

PRESSURE (CONCENTRATION) DEPENDENCE   The sorption isotherm concave to the pressure axis [Fig. 2.3(c)] is observed commonly for a penetrant gas in a glassy polymer. The isotherm is interpreted to be due to two different environments in a glassy polymer and is well described by the following equation which is composed of Henry's law and Langmuir terms.[4,5]

$$\begin{aligned} C &= C_D + C_H \\ &= k_D p + \frac{C'_H b p}{1 + bp}, \end{aligned} \tag{19}$$

where C_D is the penetrant concentration dissolved by ordinary dissolution (Henry's law mode) in the quasiliquid domains of the polymer and k_D is the

Henry's law parameter and corresponds to $S(0)$ in Eqs. (15) and (17). C_H is the penetrant concentration described by the Langmuir equation (Langmuir mode). According to Michaels et al.,[27] Langmuir adsorption occurs onto the fixed site in microvoids (the unrelaxed domains) of the glassy polymer, and C'_H and b are the maximum penetrant concentration (Langmuir capacity constant) and the affinity constant of the penetrant in the Langmuir mode, respectively. This dual-mode sorption model has been applied to many glassy-polymer–gas systems.

The plasticization of a glassy polymer is often observed in the case of glassy-polymer–vapor systems ($T < Tc$), and the sorption isotherm becomes complex like Fig. 2.3(e).[28-30] At lower pressures, it is concave and then turns convex to the pressure axis as the polymer is plasticized by the dissolved gas. Similar sorption isotherms are observed in the case of water vapor, which partly tends to cluster in a polymer. Stern and Saxena have extended the dual-mode sorption model,[31] and the solution and transport of water and organic vapors in some glassy polymers have been successfully described by the extended model.[32-34]

Another type of sorption isotherm, which is concave to the pressure axis at lower pressures and turns to a straight line that is extrapolated to the origin [Fig. 2.3(d)] has been observed for some glassy-polymer–CO_2 systems.[35,36] The difference between the two types of sorption isotherms in Figs. 2.3(d) and (e) may be due to the difference of the degree of polymer plasticization caused by the dissolved gas. Chiou and Paul[35] and Kamiya et al.[36] have extended the dual-mode sorption model by taking into account the polymer plasticization by the dissolved gas. They modified the Langmuir term to become zero at certain concentrations as a result of a depression of the glass transition temperature:

$$C'_H = C'_{H0} f(C)$$
$$f(C) = 0, \quad \text{at } C = C_g, \tag{20}$$

where C_g is called the *glass transition concentration* because the polymer isothermally transforms to the rubbery state.

Recently Stern and co-workers have presented a new model that is applicable to the two types of sorption isotherms shown in Figs. 2.3(c) and (d), but has no direct relationship to the dual-mode sorption model.[37] They derived the following relation by applying the concentration–temperature superposition principle:

$$C = S(0)p \exp\left[A\left(\frac{T_g(C)[T_g(C) - T_g(0)][T_g(0) - T]}{[T_g(0)]^2}\right)\right], \tag{21}$$

where $S(0)$ is the solubility coefficient in the limit of zero pressure, A is a parameter that depends on the temperature and the nature of the polymer–gas system, $T_g(C)$ and $T_g(0)$ are the glass transition temperatures of the polymer containing a dissolved gas at concentration C and of the pure polymer, respectively. The extension of their model to the description of diffusion and permeability coefficients is underway.[38]

Raucher and Sefcik pointed out the polymer-dissolved gas interaction and observed the polymer chain relaxation by ^{13}C nuclear magnetic resonance (NMR).[39] According to the experimental results, their "matrix" model assumes that the dissolved gas exists in a single mode in a glassy polymer and affects the gas sorption and transport properties; however, the following mathematical form is not quantitatively related to the NMR data.

$$C = \sigma_0 \exp(-\alpha C) p, \tag{22}$$

where σ_0 is the solubility in the zero-concentration limit and α is a parameter characterizing the polymer–gas interaction.

TEMPERATURE DEPENDENCE The temperature dependence of the dual-mode parameters k_D, b, and C'_H can be represented by the van't Hoff–type relations[40]

$$k_D = k_{D0} \exp(-\Delta H_D / RT) \tag{23}$$

$$b = b_0 \exp(-\Delta H_b / RT) \tag{24}$$

$$C'_H = C'_{H_0} \exp(-\Delta H^* / RT), \tag{25}$$

where k_{D0}, b_0, and C'_{H_0} are constants, ΔH_D is the difference in enthalpy of the penetrant gas dissolved by the Henry mode compared with the gas phase, and ΔH_b is the difference in enthalpy of the penetrant gas dissolved by the Langmuir mode compared with the gas phase; ΔH^* is an apparent enthalpy that characterizes the dependence of C'_H on temperature, but its physical meaning is not clear. For many glassy-polymer–gas systems, these parameters decrease with increasing temperatures; so ΔH_D, ΔH_b, and ΔH^* are usually negative.

The solubility parameters $S(0)$ in the model by Stern et al. is also an exponential function of the reciprocal absolute temperature, and the parameter A was experimentally shown to be inversely proportional to $[T_g(0) - T]$.[37]

1.4. Gas Diffusion in Polymers

When a concentration gradient of a penetrant gas is established across a nonporous polymer membrane, transport of the penetrant gas will be induced across a section in the direction of decreasing concentration. The diffusion of a penetrant gas in a polymer occurs as a result of random motions of individual molecules of the penetrant gas and the cooperative motion of polymer chain segments surrounding these molecules.

The diffusion process can be expressed phenomenologically by Fick's two laws for diffusion of a penetrants gas in a planar membrane,[6,7] as given in Eqs. (8) and (9). The diffusion coefficient D depends on the nature of the polymer–penetrant-gas system, the temperature, and generally on the penetrant concentration. When the explicit form of the concentration dependence of D is known, it is possible to solve Eq. (9), together with the pertinent initial and boundary conditions, to obtain the concentration profile $C(x, t)$ within the polymer. The diffusion flux $G(x, t)$, passing through a polymer surface at $x = l$, can be calculated from Eq. (8) and the concentration profile.

In many cases, however, the concentration dependence of D is not known, and this dependence can be elucidated from Eq. (9) and suitable experimental data. Diffusion coefficients of a penetrant gas can usually be determined from both permeability measurements and independent sorption measurements, from absorption and desorption kinetics, from spectroscopic measurements, and so on. Different experimental and computational methods of determining D have been summarized in several articles.[6-9]

1.4.1. Rubbery Polymers

PRESSURE (CONCENTRATION) DEPENDENCE In rubbery polymers, diffusion coefficients of small and less soluble penetrant gases of lower T_c, such as He, H_2, and Ne, are essentially independent of pressure (concentration) at ambient and higher temperature

$$D(C) = D(0) \quad \text{(constant)}. \tag{26}$$

The low solubility of such gases is due to their relatively weak interactions with a polymer.

On the other hand, diffusion coefficients of penetrant gases of higher T_c, such as organic vapors, are strongly dependent on concentration and can be linear or exponential functions of penetrant concentration, as shown in Figs. 2.4(a) and (b), respectively, or may be a more complex function of concentration.[9,41,42]

$$D(C) = D(0) \exp(\alpha C) \tag{27}$$
$$= D(0)(1 + \beta C), \tag{28}$$

where $D(0)$ is the diffusion coefficient in the zero-concentration limit and α and β are the constants characterizing the concentration dependence.

As the product of gas diffusivity described by Eqs. (26)–(28) and solubility described by Eqs. (15)–(17), the pressure dependences presented in Figs. 2.1(a) and (b) are commonly observed for gas permeability in rubbery polymers.

In contrast, the diffusion coefficient of water vapor in several of the less hydrophilic polymers was found to decrease, sometimes markedly, with in-

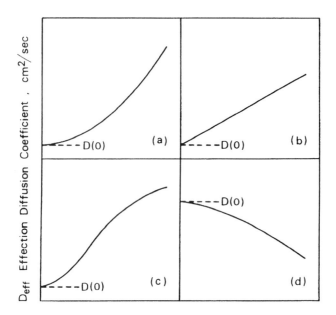

Figure 2.4 Schematic representation of typical concentration-dependent forms for diffusion coefficients. Typical for (a) vapor-rubbery polymer systems, (b) gas-rubbery polymer systems, and (c) gas-glassy polymer systems. (d) Clustering in polymers.

creasing penetrant concentration, due to clustering of water molecules [Fig. 2.4(d)].[20,34]

TEMPERATURE DEPENDENCE The mutual diffusion coefficients of many penetrant gases in rubbery polymers exhibits an exponential temperature dependence over a limited range of temperatures[4,8,21–23]

$$D = D_0 \exp(-E_d/RT), \tag{29}$$

where D_0 is a pre-exponential factor and E_d is the apparent activation energy for diffusion. Over a narrow range of temperatures, E_d is constant, and an Arrhenius plot of the logarithm of D versus $1/T$ is linear. Over wider temperature ranges, E_d decreases with increasing temperature, and the Arrhenius plot therefore shows a convex curvature. For highly soluble penetrant gases, such as organic vapors, the effective diffusion coefficient is strongly concentration dependent, and may increase or show a more complex form with increasing concentration.

1.4.2. Glassy Polymers

PRESSURE (CONCENTRATION) DEPENDENCE In glassy polymers, the measured or effective diffusion coefficient of even small penetrant gases of lower T_c is strongly concentration dependent, as shown in Fig. 2.4(c). In order to describe the transport of penetrant gases through glassy polymers, Petropoulos[43] and Paul and Koros[44] modified the "total immobilization" model,[27,45] which is associated with the dual-mode sorption model. They have assumed the penetrant species dissolved in a Henry's-law mode are fully mobile, while a part of the species dissolved in the Langmuir mode (the ratio F, $0 < F < 1$) are mobile, but not the rest of the Langmuir species. In the Paul and Koros version,[44] the total penetrant flux is expressed as a sum of two fluxes that are proportional to the concentration gradient in each mode:

$$G = D_D \frac{\partial C_D}{\partial x} - D_H \frac{\partial C_H}{\partial x}$$

$$= -D_D \frac{\partial C_D}{\partial x} - D_D \frac{\partial F C_H}{\partial x}, \tag{30}$$

where D_D and D_H are mutual diffusion coefficients for the penetrant species dissolved by the Henry's-law mode and the Langmuir mode, respectively. At a given temperature, D_D and D_H are taken to be constant. This model is called the *partial immobilization* or *dual-mobility model*.[5,44] This equation can be rewritten in terms of Fick's law with an effective diffusion coefficient D_{eff}:

$$G = -D_{eff}(C) \frac{\partial C}{\partial x}. \tag{31}$$

From Eqs. (30) and (31), the dual-mobility model expresses D_{eff} in the following form:

$$D_{eff}(C) = D_D \left(\frac{1 + FK/(1 + bp)^2}{1 + K/(1 + bp)^2} \right), \tag{32}$$

where $K = C'_H b/k_D$. This equation describes satisfactorily the concentration dependence of diffusion coefficients in many glassy polymers. As seen in Fig. 2.4(c), D_{eff} increases with concentration, approaching D_D asymptotically at high penetrant concentration. The explanation for this behavior is as follows: At low concentration (pressure), the penetrant gas dissolved in both modes contributes to the overall mobility of the penetrant in the polymer. As the penetrant concentration increases, the Langmuir sites are gradually saturated, and the more mobile Henry's-law species contribute increasingly to the overall mobility. According to the dual-mobility model, the pressure dependence of a permeability coefficient is also expressed by five parameters:

$$P = k_D D_D + \frac{C'_H b}{(1 + bp_1)(1 + bp_2)} D_H$$

$$= k_D D_D \left(1 + \frac{FK}{(1 + bp_1)(1 + bp_2)} \right). \tag{33}$$

Figure 2.1(c) illustrates the pressure dependence of the permeability coefficient described by Eq. (33).

At temperatures in the range $T < T_c$, the solubility may be sufficiently high to plasticize a glassy polymer, and the diffusion coefficient is strongly dependent on the penetrant concentration. Stern and Saxena have modified the dual-mobility model to account for an exponential concentration dependence of the diffusion coefficient of the mobile species[31] and derived the following form:

$$D_{\text{eff}} = D(0) \exp\left[\beta C_D \left(1 + \frac{FK}{1 + bp}\right)\right] \left(\frac{1 + FK/(1 + bp)^2}{1 + K/(1 + bp)^2}\right), \tag{34}$$

where β is an empirical constant characterizing the concentration dependence. The pressure dependences of the permeability coefficient as that shown in Fig. 2.1(d) is, therefore, described by the following expression:

$$P_{\text{eff}} = \frac{D(0)}{\beta p} \left\{\exp\left[\beta k_D p \left(1 + \frac{FK}{1 + bp}\right)\right] - 1\right\}. \tag{35}$$

Zhou and Stern have further extended the model to systems in which both the solubility and diffusion coefficients are concentration dependent.[46]

In the dual-mode sorption model and all modifications, the expression for the penetrant flux contains no cross terms between the two sorption modes explicitly. Fredrickson and Helfand argue that the flux expression should contain terms that account for the exchange of gas molecules between the Henry's-law mode and the Langmuir mode and have developed such an expression.[47] Barrer has also developed a model that describes gas transport in glassy polymers with coupled flows between two modes.[48]

The matrix model can also describe the diffusion coefficient and its concentration dependence by the following expression[39]

$$D(C) = D_0 \exp(\beta C), \tag{36}$$

where D_0 and β are constants and β characterizes the concentration dependence. From Eqs. (22) and (36), the pressure dependence of the permeability coefficient can be obtained as:

$$P(C) = \sigma_0 D_0 \exp(-\alpha C) \exp(\beta C). \tag{37}$$

This model needs only four parameters to describe the permeability coefficient, but further refinement seems to be necessary.[49]

2. Process Design of Gas-Separation Schemes[1]

For designing gas-separation processes it is not sufficient to know the transport properties of gases in membranes, but it is

---

[1]P_hx and P_ly in Section 2 are equal to P_1 and P_2 in Section 1, respectively.

necessary to choose proper schemes to optimize processes for practical applications.

In the following, definitions of various separation factors will first be given from transport equations of gas permeation through membranes, and then the separation factor of various stages will be discussed. Finally it will be shown that these stages are combined to give various cascades to obtain optimized separation schemes.

2.1. Separation Factor

The separation factor is an important index to express the membrane's ability for separation of a specific set of gaseous mixtures. But there are many definitions of separation factors, and the conditions that influence them are different. First their definitions will be introduced, starting from the transport equations of gases through membranes.

Let us assume that gases A and B are permeating through a membrane as shown in Fig. 2.5, from a high-pressure side (P_h) to a low-pressure side (P_l). Here gas A is permeating faster than B. A permeating flux G of gases is usually assumed to be proportional to the difference in partial pressure across a membrane with a proportionality constant P, which is called the *permeability of gas through a membrane*. The flux G is given for a binary mixture of A and B as follows:

$$G_A = P_A (P_h x - P_l y) \tag{38a}$$

$$G_B = P_B [P_h (1 - x) - P_l (1 - y)]. \tag{38b}$$

Figure 2.5 Change of mole fractions by permeation through a membrane. The gas A is concentrated in the permeate.

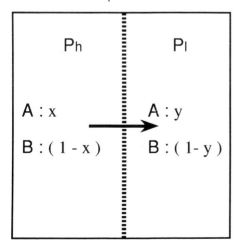

They are rewritten as follows using the pressure ratio r, which is defined as $r = P_l/P_h$, and expresses the effect of downstream pressure on separation:

$$G_A = P_A P_h(x - ry) \tag{39a}$$

$$G_B = P_B P_h[(1 - x) - r(1 - y)]. \tag{39b}$$

The ratio of these equations gives the mole fraction of the permeate y as follows:

$$\frac{G_A}{G_B} = \frac{y}{(1 - y)} = \frac{\alpha^*(x - ry)}{(1 - x) - r(1 - y)}, \tag{40}$$

where α^* is a true separation factor of a membrane, which is defined as follows and the value uniquely determined by a combination of a membrane–gases (A and B) system.

$$\alpha^* = P_A/P_B. \tag{41}$$

The most simple separation stage is shown in Fig. 2.6, where a feed F is separated to a permeate G and a nonpermeate L. The mole fraction of each is given as z, y, and x, respectively. The separation factor of this stage is defined as

$$\alpha = \frac{y/(1 - y)}{x/(1 - x)}, \tag{42}$$

which is given at a particular ratio of G to F. This ratio is called the *cut* θ and is given as

$$\theta = G/F = (z - x)/(y - z). \tag{43}$$

As is seen from Eqs. (40) and (42), α is usually smaller than α^* and becomes equal to α^* when r is equal to zero, that is, when the permeate side is under

Figure 2.6 Definition of a separation stage.

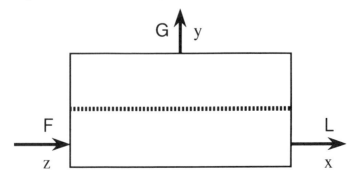

vacuum. The relation between them at cut θ is equal to zero, and is given by Blumkin[50] as follows:

$$\alpha = \frac{\alpha^*}{1 + (\alpha^* - 1)r(1 - y)/(1 - x)}, \qquad (44)$$

which shows α is not a constant as α^*, but a function of x and y. It is also not easy to estimate y from x at a given α^* using the above equation.

Theoretically it is easier to estimate x from y using a separation efficiency Z defined by Benedict, Pigford, and Levi,[51] which is given as

$$Z = \frac{y - x}{y - (x)_0}, \qquad (45)$$

where $(x)_0$ is defined as

$$(x)_0 = \frac{y}{y + \alpha^*(1 - y)}, \qquad (46)$$

which corresponds to the x value at $r = 0$, and is the smallest value of x attainable by a given membrane–gases system. By substituting y from Eq. (40) and $(x)_0$ from Eq. (45), Z is given simply as

$$Z = 1 - r, \qquad (47)$$

which expresses the effect of back pressure on separation. It is also very easy to include the effect of concentration polarization into Z. The concentration profile near a membrane separating a gas mixture is shown in Fig. 2.7, where it is seen that the mole fraction at a membrane surface x_m is lower than that in the bulk x, and their relation is given as[52]

$$\frac{y - x_m}{y - x} = \exp(J_v/k), \qquad (48)$$

where k is a mass transfer coefficient, usually defined and used in chemical engineering, and J_v is defined as

$$J_v = (G_A + G_B)/c, \qquad (49)$$

where c is the total molar concentration of a gas mixture. Then, as seen from Fig. 2.7, the total separation efficiency Z_{total} is given as follows.

$$Z_{total} = \frac{y - x}{y - (x)_0} = \frac{y - x}{y - x_m} \frac{y - x_m}{y - (x)_0}$$

$$= \exp(J_v/k)(1 - r). \qquad (50)$$

Usually calculations are performed by computer, and the effect of concentration polarization is not so large. The separation efficiency is not widely used in practical calculations, but theoretically it is a very important parameter.

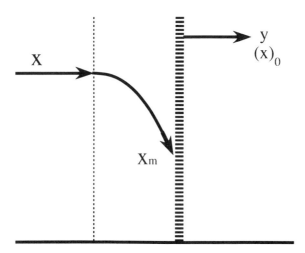

Figure 2.7 Effect of concentration polarization in gas separation.

2.2. Separation Stage

Let us go back to Fig. 2.6 and summarize separation factors. A stage separation factor α is defined by Eq. (40) as

$$\alpha = \frac{y/(1-y)}{x/(1-x)}.$$

A head separation factor β is defined as

$$\beta = \frac{y/(1-y)}{z/(1-z)}. \tag{51}$$

A tail separation factor $\bar{\beta}$ is defined as

$$\bar{\beta} = \frac{z/(1-z)}{x/(1-x)}. \tag{52}$$

Then $\alpha = \beta \bar{\beta}$.

Equation (43) was derived to obtain a stage separation factor α, defined by Eq. (42) at cut θ equal to zero. For nonzero cut cases, α values depend on the flow patterns inside stages and cannot be expressed by simple equations. To obtain these values, differential equations should be solved by computer, except for very simple cases.

There are four cases in flow patterns shown in Fig. 2.8. The most simple case is case (a), where both sides of a membrane are well mixed and x and y are constant throughout the stage. In case (b), a permeate flow is not mixed but flows vertically through a membrane. In case (c) a permeate flow concurrently with a nonpermeate, while in case (d), it flows countercurrently.

Process Design of Gas-Separation Schemes

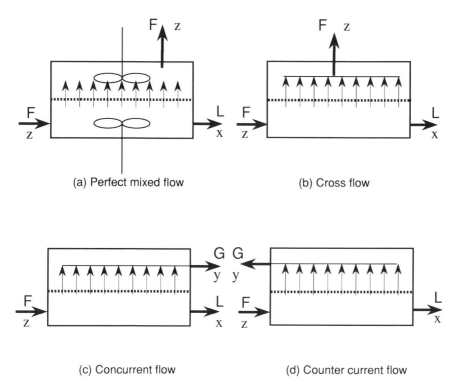

Figure 2.8 Flow schemes inside stages.

Details of the calculation methods were discussed by Blaisdell and Kammermeyer[53] and are not shown here, but some of their results are shown here. They calculated the case of oxygen enrichment from air using a silicone rubber membrane, whose separation factor α for oxygen against nitrogen is 2.05. Figure 2.9(a) gives y of oxygen in the permeate, and Fig. 2.9(b) gives x in a nonpermeate, both against cut. The countercurrent flow gives the highest y and the perfect mixed flow gives the lowest y, as shown in Fig. 2.9(a). From Fig. 2.9(b) the countercurrent flow gives the lowest x, and the perfect mixed flow gives the highest x, which means that the recovery of oxygen is best in countercurrent flow.

The reason that the countercurrent flow gives the largest separation factor is the same as that for a heat exchanger, called the *countercurrent effect* in gas-separation processes, in particular.

This treatment assumes that a true separation factor α^* is independent of x and y, and is constant. As explained in the first part of this chapter, some glassy polymers may permeate gases following the dual-sorption mode, and in this case a separation factor is not constant, but depends on concentration, and this must be taken into consideration in calculation procedures.

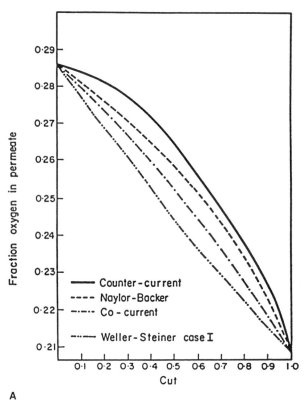

A

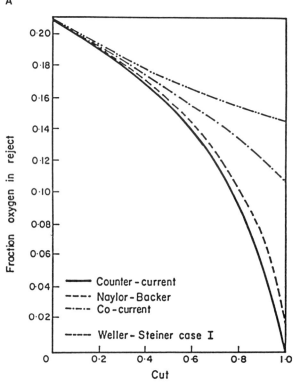

B

2.3. Separation Cascade

Figure 2.9(a) shows that oxygen concentration in the permeate cannot exceed 28.5% in a single stage, unless the separation factor of the membrane remains 2.05. In order to have a permeate with a higher oxygen concentration using such a membrane, a separation cascade should be adopted. The theory of cascades has been developed extensively for the separation of uranium isotopes,[54] but basic assumptions were very specific to this system, where the membrane separation factor was very close to 1 and concentrations of isotopes were very low. A cascade theory for more general gas-separation schemes is needed and should be developed in future, but in the meantime the general nature of cascades analyzed so far will be explained.

The scheme of a gas-separation cascade is shown in Fig. 2.10. Here a cascade consisting of f stages of enriching section and $(n - f)$ stages of stripping section is shown. The stages are so numbered that the top and bottom stages become $i = 1$ and n, respectively, i being the subscript indicating the respective stage numbers, while F_f^* and z_f^*, respectively, are the molar flow rate of the feed gas to cascade and the mole fraction of its component A. For calculating the steady-state performance of the cascade, the material balances will be given around the stage i:[55]
With respect to the total flow rate,

$$L_i + G_i = L_{i-1} + G_{i+1} + \delta_{if} F_f^*, \qquad (53)$$

where

$$L_0 = 0, \qquad G_{n+1} = 0,$$

$$\delta_{if} = \begin{cases} 1 & (i = f) \\ 0 & (i \neq f). \end{cases}$$

With respect to the flow rate of component A,

$$L_i x_i + G_i y_i = L_{i-1} x_{i-1} + G_{i+1} y_{i+1} + \delta_{if} F_f^* z_f^*. \qquad (54)$$

Equations (53) and (54) have to hold for each stage, with the flow rate and concentration profiles simultaneously satisfying these material balances.

If the concentration profile is determined, then the cut θ can be calculated from Eq. (43). Equation (53) may also be expressed in terms of the cut:

$$F_i = \bar{\theta}_{i-1} F_{i-1} + \theta_{i+1} F_{i+1} + \delta_{if} F_f^*, \qquad (55)$$

Figure 2.9 (a) Oxygen concentration versus cut for the permeated stream using the four models. $\alpha^* = 2.05$, $r = 0.359$, feed $= 0.209$ O_2 (air), 0.791 N_2. The Naylor–Backer case corresponds to the cross flow, and the Weller–Steiner case I to perfect mixed flow. (Reprinted from ref. 53, p. 1252, by courtesy of Pergamon Press, Ltd.) (b) Oxygen concentration versus cut for the reject stream. (Reprinted from Ref. 53, p. 1252, by courtesy of Pergamon Press Ltd.)

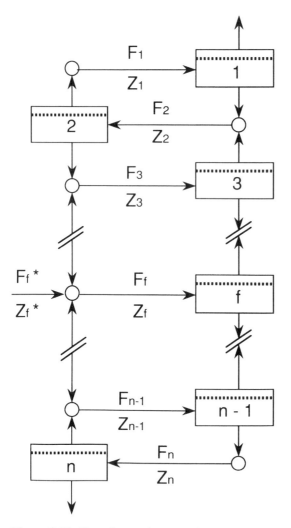

Figure 2.10 The scheme of a separation cascade.

where $\bar{\theta} = 1 - \theta$. If we set forth the material balance given by Eq. (55) for each stage, a set of linear equations is obtained with various flow rates F_i. In vector–matrix notation, these equations become

$$\mathbf{F} = \begin{bmatrix} 0 & \theta_2 & & & & \\ \bar{\theta}_1 & 0 & \theta_3 & & & \\ & \bar{\theta}_2 & 0 & \theta_4 & & \\ & & \cdots & & & \\ 0 & & & & \bar{\theta}_{n-1} & 0 \end{bmatrix} \mathbf{F} + \begin{bmatrix} 0 \\ 0 \\ F_f^* \\ 0 \\ \vdots \\ 0 \end{bmatrix}, \quad (56)$$

where $\mathbf{F} = (F_1, F_2, \ldots, F_n)^T$. Equation (56), representing a linear equation with respect to $\mathbf{F}$, can readily be solved and is given as

$$\mathbf{F} = \begin{bmatrix} 1 & -\theta_2 & & & 0 \\ -\theta_1 & 1 & -\theta_3 & & \\ & -\theta_2 & 1 & -\theta_4 & \\ & & \cdots & & \\ & & & -\theta_{n-1} & 1 \end{bmatrix}^{-1} \begin{bmatrix} 0 \\ 0 \\ F_f^* \\ 0 \\ \vdots \\ 0 \end{bmatrix} \quad (57)$$

where the matrix of cuts is all zero, except for the diagonal and the upper and lower elements. The inverse of the matrix can easily be calculated numerically.

In order to solve the material balance equations for the cascade, using the basic equations for a membrane separation cell, it is necessary to define the cascade operating conditions. There are innumerable cascade operating conditions that can meet a given exit condition. There are four theoretical forms of cascade characterized by the features prescribed in Table 2.1, in terms of the three parameters, pressure ratio r, cut θ, and separation factor α, as they may or may not vary from stage to stage. First, there is a constant cut cascade, whose cut and pressure ratio are kept constant at every stage, and to keep these conditions the separation factor should change from stage to stage. Second, there is a symmetric separation cascade, where a head and a tail separation factor of every stage is kept constant. In this case the pressure ratio and cut change from stage to stage. Third, there is a no-mixing cascade, where concentrations of feed streams to compressors should be kept the same. In this case the separation factor and pressure ratio change from stage to stage. Fourth, there is an ideal cascade that satisfies conditions in the second and the third.

In designing a membrane gas-separation cascade, the parameters first required to be determined are the total number of stages, the feed rate, the pressures, and the membrane used, that is, α^*, P_A, F_f^*, Z_f^*, for obtaining the desired products. These values are used for solving the cascade mate-

Table 2.1 Definitions of Four Forms of Cascade

		Parameters		
Designation	Characteristic Feature	Pressure Ratio	Cut	Separation Factor
Constant cut cascade	θ_i = constant	Constant	Constant	Varying
Symmetric separation cascade	$\beta_i = \bar{\beta}_i$	Constant	Varying	Varying
No-mixing cascade	$x_{i-1} = y_{i+1} = z_i$	Constant	Varying	Varying
Ideal cascade	$\beta_i = \bar{\beta}_i$, $x_{i-1} = y_{i+1} = z_i$	Varying	Varying	Constant

rial balance equations. Various method of calculations are available for this procedure, for example, methods by Hwang and Kammermeyer[56] and by Blumkin.[57] Ohno and co-workers[55] have also developed a new method. Details are not reproduced here.

2.4. New Scheme (1)

Usually a separation factor α^* of a membrane for a gas mixture is not very large; so when a certain cascade is designed by using such a membrane, the number of stages becomes large and at the same time the number of compressors becomes large, because each stage needs one compressor. This results in an excessive capital cost of the designed process, and leads to a loss of interest for engineers. To avoid such a situation, we need a better membrane having a larger separation factor, which is not easy, as discussed in the first section. In the following two methods that can improve cascade separation factors will be introduced.

2.4.1. Two-Membrane Scheme In an effort to find a way to remove radioactive krypton-85 from the nuclear fuel reprocessing procedure, Kimura et al.[58] found that the behaviors of glassy and rubbery polymers were opposite in the order of permeability values against inert gases. Silicone rubber membranes permeate larger molecular-weight gases faster, while cellulose acetate membranes permeate smaller molecular-weight gases faster. Reasons for these behaviors can be found in discussions in the first part of this chapter. When these two kinds of membranes are combined as shown in Fig. 2.11, a stage separation factor can be larger than that of each membrane. Based on this idea, stage characteristics of the two-

Figure 2.11 A two-membrane scheme. (Reprinted from Ref. 58, p. 352, by courtesy of Akademia Kiado, Acad. Sci., Budapest, Hungary.)

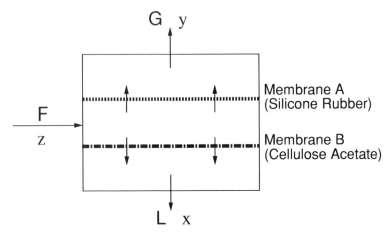

membrane scheme have been extensively studied by Ohno et al.[59] The characteristics of a silicone rubber membrane for the He-Kr system are shown in Fig. 2.12, while those of the cellulose acetate membrane are shown in Fig. 2.13. The scheme was extended further into a parallel-type separation cell shown in Fig. 2.14, where in the bottom half of the figure it is seen that cuts are selected so as to give a large stage separation factor.

2.4.2. Recycling Scheme While studying the scheme discussed previously, Ohno and co-workers[59] also found that a series-type separation cell, which is composed of only one kind of membrane, have a larger separation factor than that of a single cell. This is shown in Fig. 2.15, where it is seen that cuts are again selected to obtain a large separation factor. Table 2.2 summarizes separation factors thus obtained for the He-Kr system. A separation factor for the parallel-type separation cell, which uses silicone rubber and cellulose acetate membranes, is as large as 180, while that for the series-type cell, which uses only silicone rubber membranes, is 96. Since in practical applications it is sometimes difficult to obtain two kinds of membranes that have opposite selectivities for an objective gaseous mixture, this series-type cell may be a more realistic scheme. A comparison of separation cascades using a conventional cell and these two-unit cells was extensively studied and published elsewhere.[60]

This scheme was applied to the monitoring system of krypton-85.[61] This is shown in Fig. 2.16. The system is composed of the Kr enrichment and the Kr-Xe separation sections.

2.5. New Scheme (2)

Hwang and Thorman[62] studied on the series-type scheme, and in 1980 they proposed a "continuous membrane column (CMC)," which resembles a distillation column as shown in Fig. 2.17. It has only one compressor that produces a reflux flow. Hwang and Thorman demonstrated that by adopting this scheme the concentration of oxygen could be concentrated to 70% from air using silicone rubber membranes, whose true separation factor is 2. Since then Hwang and co-workers have been working on various aspects of CMC. Hwang and Ghalchi[63] worked on separation of CH_4 from CO_2 and/or N_2 by CMC and a combination of two columns. Yoshisato and Hwang[64] developed a new numerical technique called *orthogonal collocation* for calculating and optimizing CMC. Seok, Kang, and Hwang[65] studied CMC with two membranes (TMC). Kao, Chen, and Hwang[66] studied the effect of axial diffusion on CMC. Kothe, Chen, Kao, and Hwang[67] studied a simulation model of CMC and TMC, including axial diffusion effects, and extended it to multicomponent gas mixtures.

Stern, Perrin, and Naimon[68] compared the various schemes introduced here and discussed their merits and drawbacks. First, they compared a sin-

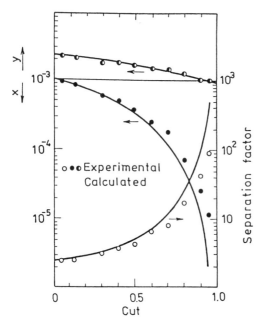

Figure 2.12 Separation of the He-Kr mixture by a silicone rubber membrane (Kr mole fraction in feed gas = 10^{-3}, high pressure = 6 atm, and low pressure = 1 atm). (Reprinted from Ref. 59, p. 301, by courtesy of Akademia Kiado, Acad. Sci. Budapest, Hungary.)

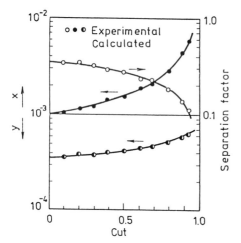

Figure 2.13 Separation of the He-Kr mixture by a cellulose acetate membrane (Kr mole fraction in feed gas = 10^{-3}, high pressure = 6 atm, and low pressure = 1 atm). (Reprinted from Ref. 59, p. 301, by courtesy of Akademia Kiado, Acad. Sci. Budapest, Hungary.)

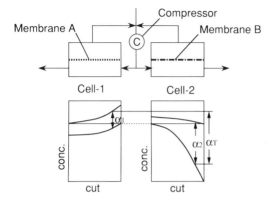

Figure 2.14 New scheme of a parallel-type separation cell. (Reprinted from Ref. 59, p. 304, by courtesy of Akademia Kiado, Acad. Sci. Budapest, Hungary.)

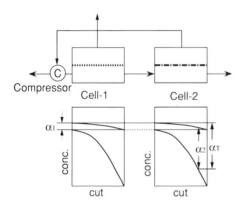

Figure 2.15 New scheme of a series-type separation cell. (Reprinted from Ref. 59, p. 304, by courtesy of Akademia Kiado, Acad. Sci. Budapest, Hungary.)

Table 2.2 Stage Separation Factors Estimated from Experimental Results for He-Kr Mixtures

Separation Cell		Membrane	Cut	Separation Factor
Conventional separation cell		Silicone rubber	0.35	3.4
Series-type separation cell	Cell 1	Silicone rubber	0.01	96
	Cell 2	Silicone rubber	0.90	
Parallel-type separation cell	Cell 1	Silicone rubber	0.90	180
	Cell 2	Cellulose acetate	0.90	

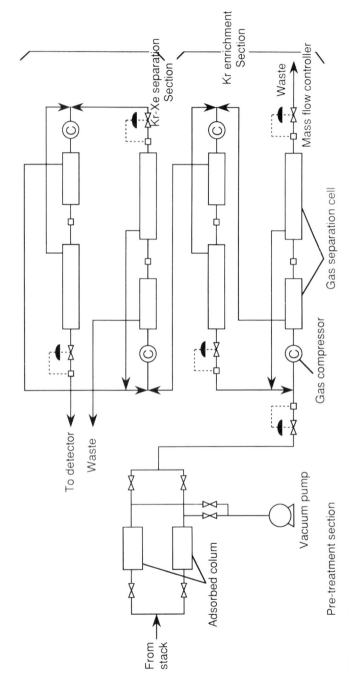

Figure 2.16 Gas treatment part flow diagram for the ^{85}Kr monitoring system.

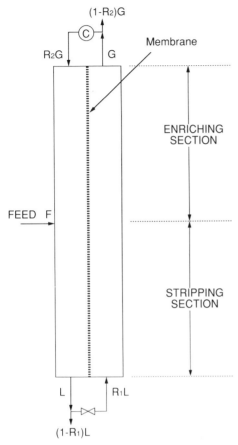

Figure 2.17 Continuous membrane column. (Reprinted from Ref. 68, p. 31, by courtesy of Elsevier Sci. Pub.)

gle countercurrent stage without recycling and that with recycling, as illustrated in Fig. 2.18. They demonstrated the effect of recycling in Fig. 2.19, where it is shown that the concentration of permeate increases with an increase of recycling fraction R and has a sharp maximum. But at the same time they pointed out an increase of compressor power and membrane area with an increase of recycling, which needs careful economic analysis for practical applications.

They also studied CMC and demonstrated the importance of recycling (reflux), as shown in Fig. 2.20, and pointed out that CMC can recover a large amount of the more permeating component.

They also demonstrated various variations of two-membrane schemes, which are shown in Fig. 2.21. Further, Perrin and Stern[69] developed mathematical models for designing two-membrane schemes, and the results

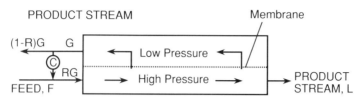

Figure 2.18 Countercurrent permeator with fractional recycling of permeate stream and high-pressure feed. (Reprinted from Ref. 68, p. 27, by courtesy of Elsevier Sci. Pub.)

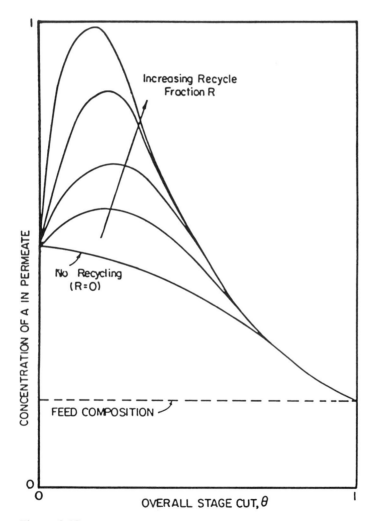

Figure 2.19 Extent of separation of a binary gas mixture on a countercurrent permeator with recycling of the permeate stream and high-pressure feed. The ordinate shows on a linear scale the concentration (in mole fraction) of the faster-permeating component of the mixture A in the permeate (low-pressure) product stream. The abscissa is also on a linear scale. (Reprinted from Ref. 68, p. 28, by courtesy of Elsevier Sci. Pub.)

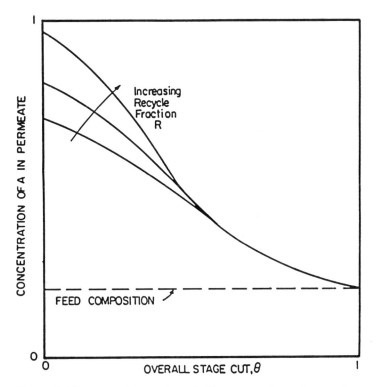

Figure 2.20 Extent of separation of a binary gas mixture in a continuous membrane column with both enriching and stripping sections. The ordinate shows on a linear scale the concentration (in mole fraction) of the faster-permeating component of the mixture A in the permeate (low-pressure) product stream. The abscissa is also on a linear scale. (Reprinted from Ref. 68, p. 34, by courtesy of Elsevier Sci. Pub.)

were confirmed experimentally[70] in the separation of He-CH$_4$ systems using silicone rubber and cellulose acetate capillary membranes.

Sengupta and Sirkar[71,72] performed a numerical analysis of the two-membrane scheme for the separation of a ternary gas mixture into three product streams, and applied it to a H$_2$-CO$_2$-N$_2$ system.

McCandless[73] used Rony's extent of separation to compare the goodness of various gas-separation schemes and extended it also to various recycling systems.[74]

So far various separation schemes have been developed, and mathematical analyses and comparisons among them have been done by various researchers. But these comparisons were done for a specific system, and their conclusions may not be applicable for general cases. Also situations and demands of various practical applications are very much different from process to process. In some cases only the permeate is important, while in other cases the recovery of components is important, and depending on

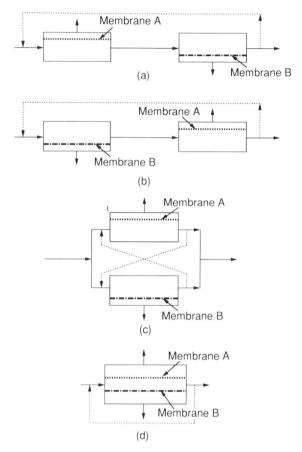

Figure 2.21 The membrane permeation system with permeators connected in series (a,b) and in parallel (c). Each permeator contains a different type of membrane. The products are both permeated (low-pressure) and unpermeated (high-pressure) streams. A two-membrane permeator (d), that is, a permeation system in which both membranes are contained in the same vessel, is shown for comparison. The dashed lines indicate recycling options. (Reprinted from Ref. 68, p. 38, by courtesy of Elsevier Sci. Pub.)

each case the optimal solution may be different. We hope that effort will be expended to reach a more general solution and to establish firm membrane technology for gas separation.

REFERENCES

1. R. D. Present, *Kinetic Theory of Gases*, McGraw-Hill, New York, 1958.
2. R. D. Present and A. J. DeBethune, *Phys. Rev.* **75**, 1050 (1949).
3. E. R. Gilliland, R. F. Baddour, and J. L. Russell, *AIChE J.* **4**, 90 (1958).

4. S. A. Stern, "Gas Permeation Process," in *Industrial Processing with Membranes*, R. E. Lacey and S. Loeb, Eds., Wiley Interscience, New York, 1972.
5. W. J. Koros and R. T. Chern, "Separation of Gaseous Mixtures Using Polymer Membranes," in *Handbook of Separation Process Technology*, R. W. Rousseau, Ed., Wiley Interscience, New York, 1987.
6. J. Crank, *The Mathematics of Diffusion*, 2nd ed., Clarendon Press, Oxford, 1975.
7. J. Crank and G. S. Park, "Methods of Measurement," in *Diffusion in Polymers*, J. Crank and G. S. Park, Eds., Academic Press, Oxford, 1968.
8. G. J. van Amerongen, *Rubb. Chem. Technol.* **37**, 1065 (1964).
9. H. L. Frisch and S. A. Stern, *Crit. Revs. Solid State and Mat. Sci.* **11**, 123 (1983), CRC Press, Boca Raton, Florida.
10. R. M. Barrer, J. A. Barrie, and N. K. Raman, *Polymer* **3**, 595; 605 (1962).
11. Y. Kamiya, T. Hirose, K. Mizoguchi, and Y. Naito, *J. Polym. Sci.: Part B: Polym. Phys.* **24**, 1525 (1986).
12. V. M. Shah, B. J. Hardy, and S. A. Stern, *J. Polym. Sci.: Part B: Polym. Phys.* **24**, 2033 (1986).
13. R. M. Barrer, *Trans. Faraday Soc.* **43**, 3 (1947).
14. C. E. Rogers, V. Stannett, and M. Szwarc, *J. Polym. Sci.* **45**, 61 (1960).
15. M. S. Suwandi and S. A. Stern, *J. Polym. Sci. Polym. Phys. Ed.* **11**, 663 (1973).
16. G. K. Fleming and W. J. Koros, *Macromolecules* **19**, 2285 (1986).
17. P. J. Flory, *Principles of Polymer Chemistry*, Cornell University, Ithaca, New York, 1969.
18. B. Katchman and A. D. McLaren, *J. Am. Chem. Soc.* **73**, 2124 (1951).
19. H. Yasuda and V. Stannett, *J. Polym. Sci.* **57**, 907 (1962).
20. J. A. Barrie, "Water in Polymers," in *Diffusion in Polymers*, J. Crank and G. S. Park, Eds., Academic Press, Oxford, 1968.
21. P. Meares, *J. Am. Chem. Soc.* **76**, 3415 (1954).
22. C. E. Rogers, J. A. Meyer, V. Stannett, and M. Szwarc, *TAPPI* **39**, 737; 741 (1956).
23. A. S. Michaels and R. B. Parker, Jr., *J. Polym. Sci.* **41**, 53 (1959).
24. S. A. Stern, J. T. Mullhaupt, and P. J. Gareis, *AIChE J.* **15**, 64 (1969).
25. L. I. Stiel and D. F. Harnish, *AIChE J.* **22**, 117 (1976).
26. M. S. Swandi, T. Hirose, and S. A. Stern, *J. Polym. Sci.: Part B: Polym. Phys.* **28**, 407 (1990).
27. A. S. Michaels, W. R. Vieth, and J. A. Barrie, *J. Appl. Phys.* **34**, 1; 13 (1963).
28. A. G. Assaf, R. H. Haas, and C. B. Purves, *J. Am. Chem. Soc.* **66**, 66 (1944).
29. A. R. Berens, *Angew. Makromol. Chem.* **47**, 97 (1975).
30. Y. Kamiya, K. Mizoguchi, T. Hirose, and Y. Naito, *J. Polym. Sci.: Part B: Polym. Phys.* **27**, 879 (1989).
31. S. A. Stern and V. Saxena, *J. Membr. Sci.* **7**, 47 (1980).
32. G. R. Mauze and S. A. Stern, *J. Membr. Sci.* **12**, 51 (1982).
33. V. Saxena and S. A. Stern, *J. Membr. Sci.* **12**, 65 (1982).
34. G. R. Mauze and S. A. Stern, *J. Membr. Sci.* **18**, 99 (1984).

35. J. S. Chiou, Y. Maeda, and D. R. Paul, *J. Appl. Polym. Sci.* **30**, 4019 (1985).
36. Y. Kamiya, K. Mizoguchi, Y. Naito, and T. Hirose, *J. Polym. Sci.: Part B: Polym. Phys.* **24**, 535 (1986).
37. Y. Mi, S. Zhou, and S. A. Stern, *Macromolecules*, **24**, 2361 (1991).
38. Personal communications with Prof. S. A. Stern and his student, Y. Mi.
39. D. Raucher and M. D. Sefcik, *Am. Chem. Soc. Sym. Ser.* **223**, 89; 111 (1983).
40. W. J. Koros, D. R. Paul, and G. S. Huvard, *Polymer* **20**, 956 (1979).
41. S. A. Stern, S. R. Sampat, and S. S. Kulkarni, *J. Polym. Sci. Polym. Phys. Ed.* **24**, 2149 (1986).
42. H. Fujita, "Organic Vapors above the Glass Transition Temperature," in *Diffusion in Polymers*, J. Crank and G. S. Park, Eds., Academic Press, Oxford, 1968.
43. J. H. Petropoulos, *J. Polym. Sci.; Part A2* **8**, 1797 (1970).
44. D. R. Paul and W. J. Koros, *J. Polym. Sci., Polym. Phys. Ed.* **14**, 675 (1976).
45. D. R. Paul, *J. Polym. Sci., Part A2* **7**, 1811 (1969).
46. S. Zhou and S. A. Stern, *J. Polym. Sci.: Part B: Polym. Phys.* **27**, 205 (1989).
47. G. H. Fredrickson and E. Helfand, *Macromolecules* **18**, 2201 (1985).
48. R. M. Barrer, *J. Membr. Sci.* **18**, 25 (1984).
49. T. A. Barbari, W. J. Koros, and D. R. Paul, *J. Polym. Sci.: Part B: Polym. Phys.* **26**, 709; 729 (1988).
50. S. Blumkin, Oak Ridge National Laboratory Report, K-OA-1559, Jan. 15 (1968).
51. M. Benedict, T. Pigford, and H. Levi, *Nuclear Chemical Engineering*, 2nd ed., p. 825, McGraw-Hill, New York, 1981.
52. O. Bilous and G. Counas, 2nd Geneva Atomic Conf., P/1263 (1958).
53. C. T. Blaisdell and K. Kammermeyer, *Chem. Eng. Sci.* **28**, 1249 (1973).
54. M. Benedict, T. Pigford, and H. Levi, *Nuclear Chemical Engineering*, 2nd ed., p. 651, McGraw-Hill, New York, 1981.
55. M. Ohno, T. Morisue, O. Ozaki, and T. Miyauchi, *J. Nucl. Sci. Tech.* **15**, 411 (1978).
56. S. T. Hwang and K. Kammermeyer, *Can. J. Chem. Eng.* **43**, 36 (1965).
57. S. Blumkin, Oak Ridge National Laboratory Report, K-OA-1559 (1968).
58. S. Kimura, T. Nomura, T. Miyauchi, and M. Ohno, *Radiochem. Radioanal. Lett.* **13**, 349 (1973).
59. M. Ohno, T. Morisue, O. Ozaki, H. Heki, and T. Miyauchi, *Radiochem. Radioanal. Lett.* **27**, 299 (1976).
60. M. Ohno, T. Morisue, O. Ozaki, and T. Miyauchi, *J. Nucl. Sci. Tech.* **15**, 376 (1978).
61. S. Takamatsu, M. Ohno, T. Miyazawa, and O. Ozaki, Int. Symp. on the Monitoring of Radioactive Airborne and Liquid Release from Nuclear Facilities, IAEA-SM 217/7 (1977).
62. S. T. Hwang and J. M. Thorman, *AIChE J.* **26**, 558 (1980).
63. S. T. Hwang and S. Ghalchi, *J. Memb. Sci.* **11**, 187 (1982).
64. R. A. Yoshisato and S. T. Hwang, *J. Memb. Sci.* **18**, 241 (1984).

65. D. R. Seok, S. G. Kang, and S. T. Hwang, *J. Memb. Sci.* **27**, 1 (1986).
66. Y. K. Kao, S. Chen, and S. T. Hwang, *J. Memb. Sci.* **32**, 139 (1987).
67. K. D. Kothe, S. Chen, Y. K. Kao, and S. T. Hwang, *J. Memb. Sci.* **46**, 261 (1989).
68. S. A. Stern, J. E. Perrin, and E. J. Naimon, *J. Memb. Sci.* **20**, 25 (1984).
69. J. E. Perrin and S. A. Stern, *AIChE J.* **31**, 1167 (1985).
70. J. E. Perrin and S. A. Stern, *AIChE J.* **32**, 1889 (1986).
71. A. Sengupta and K. K. Sirkar, *J. Memb. Sci.* **21**, 73 (1984).
72. A. Sengupta and K. K. Sirkar, *J. Memb. Sci.* **39**, 61 (1988).
73. F. P. McCandless, *J. Memb. Sci.* **19**, 101 (1984).
74. F. P. McCandless, *J. Memb. Sci.* **24**, 15 (1985).

3 Relationships between Aggregation State and Gas Permeation Properties

Tisato Kajiyama

1. Relationships of Surface Structure to Gas Permeation in Polymer/Molecular Membrane Composite Films
   1.1. Aggregation Structure of Composite Thin Films
   1.2. Surface Chemical Composition of Composite Thin Films
   1.3. Relationships of Surface Chemical Composition to Gas Permselectivity in Polymer/Mixed Multibilayer Membrane Composite Films
2. Relationships of Molecular Aggregation to Gas Permeation in Polymers Composed of Rigid and Semirigid Sequences
   2.1. Molecular Characterization and Glassy State of Poly[spiro(2, 4) hepta-4,6-diene]
   2.2. Aggregation State of PSHD Chains
   2.3. Relationships of Molecular Packing to Gas Permeation in PSHD
3. Phase Transition and Molecular Orientation Effects on Gas Separation through Polymer/LC Composite Films
   3.1. Construction and Aggregation State of Polymer/LC Composites
   3.2. Phase Transition Effects on Gas Permeation through Polymer/LC Composite Membranes
   3.3. Molecular Filtration of Hydrocarbon Isomers through Polymer/LC Composite Membranes Based on Molecular Orientation
   3.4. Oxygen Enrichment through Polymer/LC/Fluorocarbon Monomer Ternary Composite Membranes

The permeation of gases through a polymeric membrane depends on the sorption and diffusion processes of the gas molecules. The factors affecting it (i.e., solubility and diffusivity) have been correlated with the chemical structure and physical characteristics of the constituent molecules of the membranes.[1] For instance, the factors controlling physical properties, such as the glass transition behavior,[2,3] molecular motions[4,5] the degrees of crytallinity,[6] crosslinking,[7,8] and grafting[9] affect the diffusivity of gases in the polymers. On the other hand, the factors controlling both chemical and physical properties such as plasticizer[10,11] and substituent groups[12] also affect the solubility of gases in the polymeric membranes.

1. Relationships of Surface Structure to Gas Permeation in Polymer/Molecular Membrane Composite Films

The structure and properties of Langmuir–Blodgett (LB or builtup) films have received much attention recently. This is due to the fact that the control of the state of molecular aggregation is easily achieved by the preparation technique developed by Langmuir[13] and Blodgett.[14] Application studies concerning electron conduction, semiconductors, biomaterials, etc. have been carried out by several authors.[15–17] X-ray photoelectron spectroscopy (XPS) is a usesful technique to characterize the surface composition of solids. The estimation of the photoelectron mean free path in organic solids has been achieved by utilizing builtup films.[18–20] Since the builtup film forms a highly ordered layer structure, it is a suitable material for the estimation of the photoelectron mean free path. The electron mean free path of the C_{1s} photoelectron (kinetic energy ~ 970 eV) in the builtup films of long-chain fatty acids ranges from 4 to 5 nm.[20] A series of amphiphiles containing fluoroalkyl chains has been synthesized by Kunitake and co-workers.[21,22] Fluorocarbon amphiphiles form a bilayer structure in water and show phase transition behavior similar to that observed for biological lipids or artificial hydrocarbon amphiphiles.[23–25] The fluorocarbon amphiphiles can be immobilized by casting an aqueous solution of amphiphile with aqueous poly(vinyl alcohol).[26] The fluorocarbon amphiphile molecules in this composite thin film form highly oriented lamellae similar to those of builtup films.[27–29] Multicomponent bilayers of double-chain ammonium amphiphiles were immobilized on Millipore membranes as the composite film with poly(vinyl alcohol).

1.1. Aggregation Structure of Composite Thin Films

The fluorocarbon amphiphile and polymer used in this study are shown in Fig. 3.1. The fluorocarbon amphiphile forms a stable bilayer membrane in water. The fluorocarbon amphiphile was dispersed in

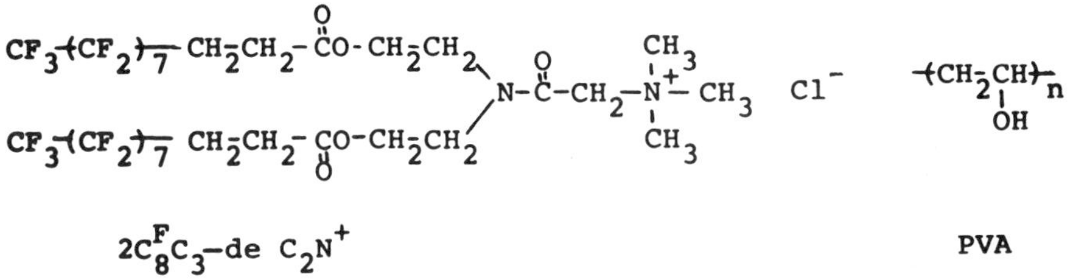

Relationships of Surface Structure to Gas Permeation

$$CF_3\text{-}(CF_2)_7\text{-}CH_2CH_2\text{-}\overset{O}{\overset{\|}{C}}O\text{-}CH_2CH_2$$
$$CF_3\text{-}(CF_2)_7\text{-}CH_2CH_2\text{-}\overset{O}{\overset{\|}{C}}O\text{-}CH_2CH_2$$
$$N\text{-}\overset{O}{\overset{\|}{C}}\text{-}CH_2\text{-}\overset{CH_3}{\overset{|}{\underset{|}{N^+}}}\text{-}CH_3 \quad Cl^-$$

$$2C_8^FC_3\text{-de } C_2N^+$$

$$\text{-}(CH_2\underset{|}{CH})_n\text{-}$$
$$\qquad\quad OH$$

PVA

Figure 3.1 Chemical structures of fluorocarbon amphiphile and poly (vinyl alcohol) (PVA).

water by sonication, and the dispersion was mixed with a water solution of poly(vinyl alcohol) (PVA) (MW = 154,000). The composite thin film was prepared by casting a water solution of PVA with the fluorocarbon amphiphile on a clean glass plate at room temperature, and the cast film was then extensively dried in vacuo. The weight fraction of the fluorocarbon amphiphile in the composite thin film was varied from 19 to 83 wt %. The aggregation state of amphiphile molecules in the composite thin film was studied by wide-angle x-ray diffraction. Figure 3.2 shows the wide-angle x-ray diffraction (XD) pattern of a fluorocarbon amphiphile powder at room temperature. Since the phase transition temperature of the fluorocarbon amphiphile is 351 K, this amphiphile is in a crystalline state at room temperature.[29] The fluorocarbon amphiphile shows x-ray reflections corresponding to 4.33, 2.24, 1.48, and 1.11 nm spacings. As the reciprocal spacings from the fluorocarbon amphiphile are in the ratio of 1:2:3:4, this fluorocarbon amphiphile aggregates in a bimolecular lamellar form.[29,30] The wide-angle x-ray reflections of 0.46, 0.49, and 0.56 nm were observed in the case of a crystalline state of fluorocarbon amphiphile. These reflections became a halo above the crystal–liquid-crystal phase transition temperature T_c. This indicates that these reflections are related to the melting of fluoroalkyl chains in the fluorocarbon bilayer lamella. Figure 3.3 shows XD patterns for the composite thin films containing the fluorocarbon amphiphile of 19 and 83 wt %. When the x-ray beam was irradiated perpendicular to the film surface (through view, TV), two Debye rings corresponding to intermolecular spacings of 0.46 and 0.49 nm were observed. However, the TV diffraction pattern did not show any reflection related to the lamellar structure. When the x-ray beam was irradiated parallel to the film surface (edge view, EV), the XD patterns showed the orientation of bimolecular lamellae perpendicular to the film surface. On the meridian of EV, the higher-order reflections attributed to the bilayer lamellar structure were observed. This result indicates that the fluorocarbon amphiphile forms bilayer lamellae parallel to the surface of the composite thin film. The intensity of these reflections became more distinct with an increase in the weight fraction of fluorocarbon amphiphile. The absence of these

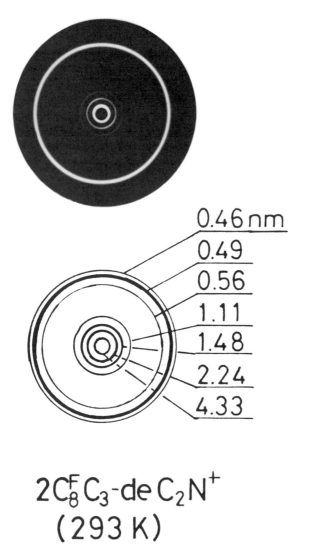

Figure 3.2 X-ray diffraction pattern of fluorocarbon amphiphile (powder) at 300 K.

small-angle reflections on TV supports that the orientation of the bilayer lamellae is almost parallel to the film surface.

The composite thin film was fractured in liquid nitrogen in order to observe its inner morphology with a scanning electron microscope. Since the liquid nitrogen temperature is much lower than the glass transition temperature of PVA and the phase transition temperature of the fluorocarbon amphiphile, the fracture surface clearly exhibits its internal structure without artifacts. Figure 3.4 shows a scanning electron micrograph of

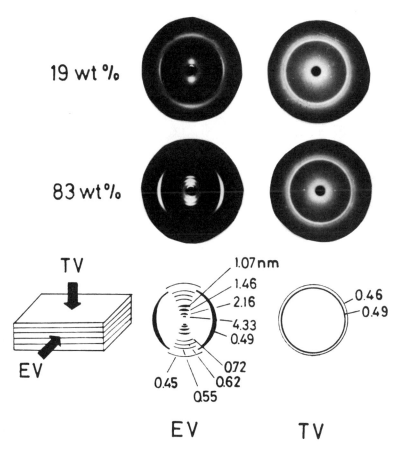

Figure 3.3 Wide-angle throughview (TV) and edge view (EV) x-ray diffraction patterns for the composite thin films containing fluorocarbon amphiphile at 19 and 83 wt %.

the composite thin film containing 83 wt % fluorocarbon amphiphile. Although PVA homopolymer showed the smooth surface characteristic of brittle failure, distinct striations parallel to the film surface were observed inside the composite thin film. As the x-ray diffraction study revealed the orientation of bimolecular lamellae parallel to the film surface, these striations may be related to the aggregation of the bimolecular lamellae of amphiphile molecules in the composite thin film. These striations became less distinct with a decrease in the weight fraction of the fluorocarbon amphiphile. Polarized optical microscopic observation was carried out to investigate the orientation of the amphiphile molecules in the composite thin film. The incident light was perpendicular to the surface of the composite thin films. The PVA film and the composite thin film used in this study showed a dark field under crossed nicols. This indicates that the orienta-

Figure 3.4 Scanning electron micrograph of the fracture surface of the composite thin films containing fluorocarbon amphiphile at 83 wt %.

tion of the amphiphile molecular axes is nearly perpendicular to the film surface. This result is consistent with the x-ray diffraction results as shown in Fig. 3.3.

1.2. Surface Chemical Composition of Composite Thin Films

The surface structure of composite thin films was evaluated by means of XPS. Figure 3.5 shows the XPS spectra of C 1s for the air-facing surface (AFS) of the composite thin film containing 19 wt % fluo-

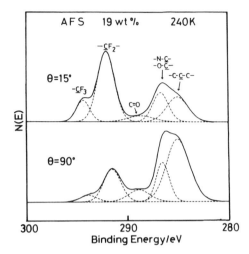

Figure 3.5 C_{1s} spectra for the air-facing surface (AFS) of the composite thin film containing 19 wt % fluorocarbon amphiphile at takeoff angles of 15° and 90°.

rocarbon amphiphile at the takeoff angles θ of 90° and 15°. The C_{1s} peak observed at 286.3 eV is mainly contributed by the ether carbon in PVA. However, the C_{1s} peaks at 292.1 and 294.4 eV are assigned to the fluorocarbon groups. Thus the relative intensity of C_{1s} from fluorocarbon to the total C_{1s} corresponds to the fraction of fluorocarbon on the surface. The relative intensity of fluorocarbon to the total C_{1s} for a takeoff angle of 15° was greater than that for a takeoff angle of 90°. The analytical depth in the case of the takeoff angle of 15° is one-fourth that of a takeoff angle of 90°.[29] This indicates that the concentration of fluorocarbon on the outermost surface is larger than that of the inner one.

Figure 3.6 shows the variation of the fraction of fluorocarbon in the total C_{1s} for AFS of the composite thin film with the weight fraction of fluorocarbon amphiphile. The "bulk" indicates the fraction of fluorocarbon in total carbon calculated from the elemental composition of each specimen. For a takeoff angle of 90°, the relative intensity of fluorocarbon increased with the fraction of fluorocarbon amphiphile in the bulk. The relative intensity of fluorocarbon in the lower fraction region of fluorocarbon amphiphile was larger than the calculated "bulk" value. This is due to the enrichment of fluorocarbon amphiphile on AFS. This occurs because of the extremely low surface free energy of the fluorocarbon groups in the amphiphile. With an increase in the weight fraction of amphiphile, the relative intensity of fluorocarbon for the takeoff angle of 90° approaches the bulk value. The relative intensity of fluorocarbon for the takeoff angle of 15° is larger than that of 100% fluorocarbon amphiphile. This may be ascribed to the orientation of fluorocarbon chains to the interface between air and the composite thin film.

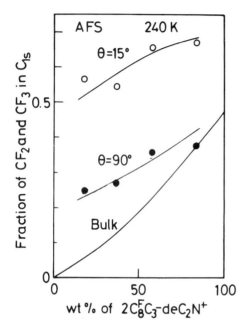

Figure 3.6 Variation of the fraction of fluorocarbon in total C_{1s} on the AFS of the composite thin film with weight percent of fluorocarbon amphiphile at takeoff angles of 15° and 90°.

For the quantitative analysis, a simple layer model was applied to the analysis of the surface composition of this composite thin film.[31] Figure 3.7 shows the patchy overlayer model for the composite thin film; P and F denote poly(vinyl alcohol) (PVA) and bimolecular membrane of fluorocarbon amphiphile, respectively. t in Fig. 3.7 corresponds to the layer thickness of fluorocarbon amphiphile in the composite thin film. Since the fluorocarbon amphiphiles form bimolecular lamellae, t is the magnitude of bilayer thickness, 4.4 nm. For simplicity, it is assumed that the concentration of each atom functional group is constant in each layer. If we assign N_F, N_B, and N_P as the atomic percent of fluorocarbon in the F layer, of carbon atoms except fluorocarbon in the F layer, and of total carbon in the P layer, the contribution of each carbon atom to the C_{1s} peak is expressed by

$$I_F \propto N_F G[1 - \exp(-t/\lambda_C \sin \theta)] \tag{1}$$

$$I_B \propto N_B G[1 - \exp(-t/\lambda_C \sin \theta)] \tag{2}$$

$$I_P \propto N_P(1 - G + G \exp(-t/\lambda_C \sin \theta)) \tag{3}$$

where I_F, I_B, and I_P are the intensities from fluorocarbon in the F layer, from carbon atoms except for fluorocarbon in the F layer, and from total carbon in the P layer. G is a variable that indicates the fraction of the sur-

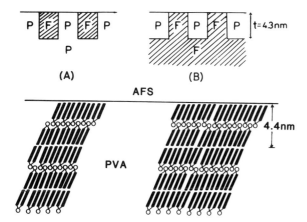

Figure 3.7 Patchy overlayer model for the composite thin film. F and P are the bimolecular lamella of fluorocarbon amphiphile and the poly (vinyl alcohol) layer, respectively.

face area covered with bimolecular lamellae of fluorocarbon amphiphile. The magnitude of λ_C employed was 4 nm and was obtained for a builtup film of fluorocarbon amphiphile.[31] The fraction of fluorocarbon in the total C_{1s} is expressed as

$$X_F = I_F/(I_F + I_B + I_P). \tag{4}$$

The actual calculation was carried out by expanding the model up to the seventh layer. Another calculation was carried out for the layer model in which PVA layers exist as overlayers of bimolecular lamellae of fluorocarbon amphiphile [Fig. 3.7(b)]. Since the fluorocarbon group shows low surface free energy compared with that of PVA, the model with fluorocarbon bilayer as an overlayer [Fig. 3.7(a)] has been employed.

Figure 3.8 shows the variation of the fraction of fluorocarbon in total C_{1s} with takeoff angle of the photoelectron as a function of the fraction of surface coverage G with bilayer lamellae of fluorocarbon amphiphile. The number of continuous bilayers on the outermost layer in Fig. 3.8 is 0.5 (a) and 3.5 (b). The calculated curve in the case that the surface is completely covered with a bimolecular lamella of fluorocarbon amphiphile for the takeoff angles of 15° and 90° agreed fairly well with the observed value for the composite thin film with 19 wt % of fluorocarbon amphiphile. This indicates that even for a small bulk fraction of amphiphile, surface coverage with fluorocarbon amphiphile occurs due to the small surface free energy of the fluoroalkyl group. The calculated curve for complete coverage with more than four continuous layers agreed fairly well with the observed value for the composite thin film with 83 wt % fluorocarbon amphiphile. Figure 3.9 exhibits the variation of the fraction of fluorocarbon to the total

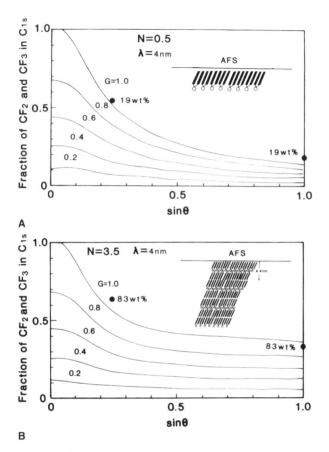

Figure 3.8 Variation of the calculated fraction of fluorocarbon in total C_{1s} for the AFS of the composite thin film with takeoff angle as a function of the surface coverage ratio G based on the patchy overlayer model. The number of bilayers N on the outermost surface is (a) 0.5 and (b) 3.5, respectively. Closed circles show the observed fraction of fluorocarbon for the composite thin film.

carbon with the number of bilayers at the outermost surface in the case of complete surface coverage with fluorocarbon amphiphile bilayer. The comparison between the observed fraction and the calculated one at takeoff angles of 15° and 90° gives the number of repeating bilayers on the surface. Table 3.1 summarizes the apparent magnitude of the number of bilayers on the surface of the composite thin film. Since PVA is a hydrophilic polymer, the fluorocarbon amphiphiles orient the hydrophilic group to PVA at their interfacial parts between PVA and amphiphile.

The surface molecular aggregation structure of the composite thin film is schematically represented in Fig. 3.10. In the case of a composite thin film with a small fraction of amphiphile, the whole film surface is com-

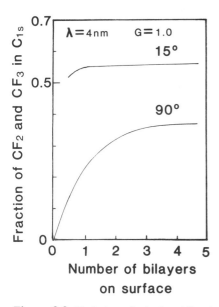

Figure 3.9 Variation of calculated fraction of fluorocarbon in total C_{1s} with the number of bilayers on the outermost surface at takeoff angles of 15° and 90°.

Table 3.1 Number of Bilayers on the Surface of the Polymer/Fluorocarbon Amphiphile Composite Thin Film

Wt% of Amphiphile	Number of Bilayers
19	0.9
38	1.5
58	4.0
83	4.0

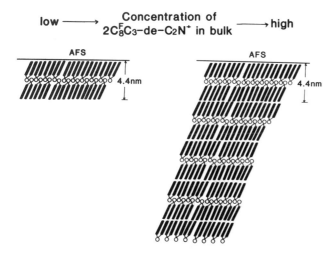

Figure 3.10 Schematic representation of the structure of the outermost layer of the composite thin film with small and large weight fractions of fluorocarbon amphiphile.

pletely covered with monolayer or bimolecular lamellae of fluorocarbon amphiphilic molecules. With an increase in the fraction of fluorocarbon amphiphile in the composite thin film, the thickness of the bilayer that orients its fluoroalkyl group to the air increased with the fraction of fluorocarbon amphiphile in the bulk. The surface of the composite thin film with 83 wt % fluorocarbon amphiphile is completely covered with continuous bilayers.[31] It has been reported that permselectivity of oxygen to nitrogen gases α of this composite thin film showed a large magnitude even if the fraction of fluorocarbon amphiphile was very small. This surface structure analysis revealed that this increase in α is due to the enrichment of fluorocarbon amphiphile on the outermost surface of the composite thin film. This result will be discussed in more detail in the next section.

1.3. Relationships of Surface Chemical Composition to Gas Permselectivity in Polymer/(Mixed Multibilayer Membrane) Composite Films

In general, permeability coefficients of the bilayer membrane are enhanced when the bilayer membrane is in the liquid-crystalline rather than the crystalline phase. Fluorocarbon bilayer films show enhanced oxygen permeation due to the affinity between fluorocarbon moieties and molecular oxygen. In contrast, fluorocarbon bilayers possess high phase transition temperatures (T_c), and the enhanced permeation due to the liquid-crystalline state cannot be achieved at ambient temperatures. We therefore intended to combine oxygen affinity of the fluorocarbon bilayer and enhanced permeation through the liquid-crystalline hydrocarbon bilayer. Fluorocarbon and hydrocarbon bilayers are not readily miscible,[32,33] and their composite bilayers may exhibit advantages of both of the component bilayers. The cast films of multicomponent bilayers of the amphiphiles shown in Fig. 3.11 were prepared in order to examine their component distribution and gas permeation characteristics.

Figure 3.12 exhibits the temperature dependences of the permeability coefficients (P_{O_2} and P_{N_2}) and the separation factor ($\alpha = P_{O_2}/P_{N_2}$) for the (2F/2H)/PVA composite films. At temperatures below T_c of the hydrocarbon component (2H), the P values are not affected by the content of the fluorocarbon component, being in the range of 10^{-9}–10^{-10} cm^3 (STP) cm^{-1} · s^{-1} (cm Hg^{-1}). The permeation data for simpler composite films at 30°C are also given in the figure. In the case of the 2F (70 wt %)/PVA film, both the P_{O_2} and P_{N_2} value are close to those of the (2F/2H)/PVA film, and the values for the 2H(50 wt %)/PVA film (uniform films were not obtainable at higher 2H content) are smaller roughly by an order of magnitude. Gas permeability is undoubtedly enhanced by the presence of the 2F layer.

The P values for the (2F/2H)/PVA film are greatly enhanced in the T_c region and level off at higher temperatures. The extent of the drastic in-

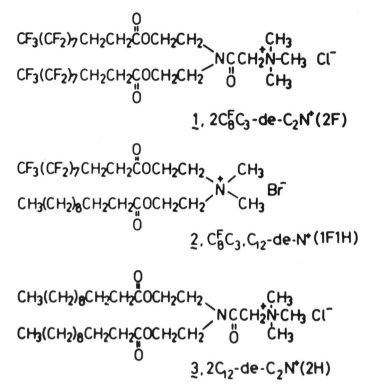

Figure 3.11 Chemical structures of double fluorocarbon (2F), fluorocarbon–hydrocarbon (1F1H), and double hydrocarbon (2H) amphiphiles.

crease is enhanced with increasing content of the hydrocarbon component: ~ 20 times enhancement in P for a cast film with 2F:2H = 1:2. The enhanced fluidity of the 2H bilayer domain is responsible for the P jump, since the 2F component does not show phase transition behavior in these temperature ranges. Unfortunately, a homogeneous cast film could not be obtained at higher 2H contents, and further enhancement in P at $T > T_c$ was not attained in the (2F/2H)/PVA system.

The separation factor α changes little with changes in the component composition. Maximal values (α = 2.7–2.9) are attained at the higher end of the T_c region and become smaller at higher and lower temperatures. The α values for the 2F/PVA film and the 2H/PVA film are 2.3 and 1.3, respectively, at 30°C. It is clear that the large values observed for the (2F/2H)/PVA films arise from the presence of the 2F layer.

A wider variation of the component composition was possible with the 2F/2H/1F1H composite films. Figure 3.13 gives the temperature dependence of P for polymer/(multicomponent bilayers) composite films. The P values (for O_2 and N_2) are virtually unaffected by the component compo-

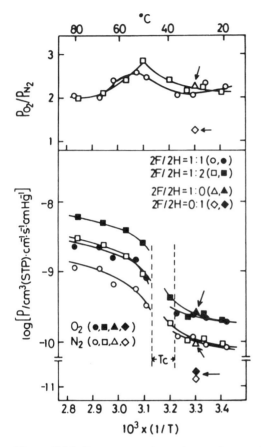

Figure 3.12 Temperature dependence of permeability coefficient (P) and the ratio P_{O_2}/P_{N_2} for the 2F/2H-PVA films. Data for 2F-PVA and 2H-PVA films are denoted by arrows.

sition at temperatures below T_c of the 2H bilayer, but are greatly enhanced at the higher end of the T_c region (indicated by the two dotted lines in Fig. 3.13). The extent of the P jump again increases with increasing content of the hydrocarbon component, being ~ 100 times for the PVA composite film of 2F/2H/1F1H (1:7:1). The extent of the P jump is reduced to 10 times when the amount of the hydrocarbon component is decreased (2F:2H:1F1H = 1:1:1). The permeability experiment was also conducted for the 1F1H (70 wt %)/PVA film. The log P value linearly decreased with $1/T$, α being 1.3–1.4. These results clearly show that gas permeation is promoted by the presence of the fluid, liquid-crystalline phase of the 2H bilayer. The separation factor again showed a maximum at the higher end of T_c and was insensitive to the bilayer composition.[27,28]

We carried out the permeation experiment by using composite mem-

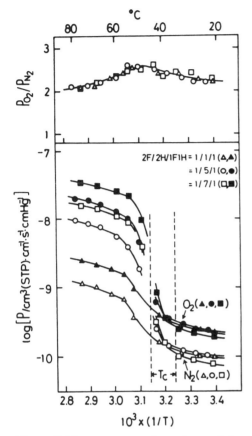

Figure 3.13 Temperature dependence of the permeability coefficient P and the ratio P_{O_2}/P_{N_2} for the 2F/2H/1F1H-PVA films.

branes composed of polymer and thermotropic nematic liquid crystal.[4,34,35] The liquid crystal forms continuous phases in the polymer matrix, and the P value is enhanced by factors of 100–500 by a temperature change of a few degrees near T_c of the liquid crystal. The viscosity of the nematic phase is in the same range as that of organic solvent or water, and the liquid-crystalline domain supposedly acts as the efficient mobile region of the gases. It was concluded that the temperature dependence of P is influenced by the thermal motion of the membrane component as well as by the continuity and/or size of the liquid-crystal phase. An analogous argument may be presented for the interpretation of the permeation characteristics of the multicomponent bilayer film. The discontinuous jump of P at T_c is ascribable to greater gas mobility in the liquid-crystalline phase of the hydrocarbon component, and the jump becomes larger as the hydrocarbon domain is enlarged.

2. Relationships of Molecular Aggregation to Gas Permeation in Polymers Composed of Rigid and Semirigid Sequences

Poly[spiro(2,4)hepta-4,6-diene] (PSHD) is an amorphous random copolymer composed of rigid sequences of 1,4 units and semirigid sequences of 1,2 units. Its chemical structure is

[Chemical scheme: SHD →(Ph₃CBF₄) 1,4-unit + 1,2-unit (PSHD)]

The characteristics of the 1,4 and 1,2 unit sequences were determiend from viscoelastic measurements and rotational energy calculations. The term *semirigid sequence* means that the potential barrier between two minima on rotation about the 1,2–1,2 bond is much lower than those for the 1,4–1,4 and 1,4–1,2 bonds and also that the 1,2–1,2 bond can be twisted only in a narrow range of rotational angle between these two minima. The density of PSHD is considerably lower than those of typical flexible polymers and decreases by increasing the fraction of the 1,4 unit, $F_{1,4}$, because the rigid sequences connected by the semirigid joints of the 1,2–1,2 bonds cannot pack themselves densely and in parallel. Conformations that can be taken by the molecular chains are considerably restricted in comparison with those of completely flexible chains. If PSHD were entirely composed of the 1,4 unit, its chains would align in parallel and crystallize without folding. Unfortunately, the synthesis of 1,4-PSHD has not as yet been successful under various polymerization conditions. However, it might be interesting to compare the amorphous structure of PSHD composed of the rigid and semirigid sequences with those of flexible polymers, since the aggregation states of PSHD and ordinary flexible polymers must be quite different.

Therefore we suggest that it may be possible to prepare the PSHD film with characteristics of a permselective membrane for gases or liquids by controlling the inhomogeneity of the amorphous state.[36] In this section, the aggregation states of the PSHD molecules have been investigated by x-ray diffraction and electron microscopy. The relation of the permeabilities and diffusivities of gases to the structural characteristics associated with rigid sequences jointed by semirigid sequences is also discussed.

2.1. Molecular Characterization and Glassy State of Poly[spiro(2,4)hepta-4,6-diene]

Samples of poly[spiro(2,4)hepta-4,6-diene] (PSHD) were prepared by cationic polymerization of spiro[2,4]hepta-4,6-diene (SHD) using various intiators in the manner described by Ohara et al.[37] The fraction of 1,4 units $F_{1,4}$ was determined from the relative peak areas of NMR spectra for olefinic protons of the 1,2 and 1,4 units.[38] Thus, $F_{1,4}$ ranges from 0.31 to 0.85 depending on the kinds of initiator, the solvent, and the polymerization temperature. The maximum $F_{1,4}$ was found to be 0.85 in this study, and it appears that PSHD with $F_{1,4} = 1.0$ cannot be prepared by cationic polymerization. The fraction of 1,4 units ranges from 0.65 to 0.72, and seems to decrease slightly with decreasing number-average molecular weight or [η]. This suggests that the tendency for sequential propagation of 1,2 units is less than that of 1,4 units, since a globular conformation is presumably formed by sequential addition of 1,2 units.

Figure 3.14 shows the relation between [η] and $\bar{M}_n$. The Sakurada–Mark–Houwink viscosity equation is

$$[\eta] = 6.0 \times 10^{-6} \bar{M}_n^{0.81}, \qquad \text{m}^3/\text{kg, 313 K, THF.} \tag{5}$$

Though the solvent and temperature are different from those used by Ohara et al.,[37] the exponent 0.81 is unexpectedly far smaller than the value (1.71) they reported, and is comparable with that of ordinary flexible-chain polymers in good solvents. The geometries of the dimeric model compounds were determined by reference to cyclopentene[39] and SHD,[40] as shown in Fig. 3.15. The five-membered ring was assumed to lie in a plane. Potential energy calculations were performed only for the nonbonded interactions. Electrostatic energy was ignored in evaluating the potential energy because there are no polar or ionic atomic groups. The potential function used for calculation of the nonbonded energy was of the Lennard–Jones type, as shown in the following eq.:

$$U_{nb}(r_{ij}) = FA/r_{ij}^{12} - C/r_{ij}^{6}, \tag{6}$$

where r_{ij} is the distance between the ith and jth atoms. The values of the coefficients A and C used were those of Momany et al.[41] The factor F was

Figure 3.14 Plot of [η] against $\bar{M}_n$ for fractionated PSHDs. Viscosity in tetrahydrofuran solutions was measured at 313 K.

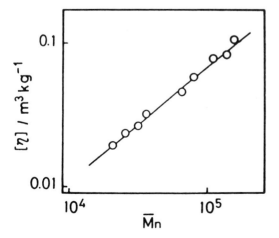

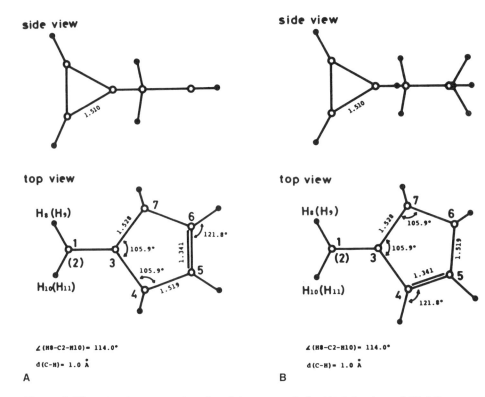

Figure 3.15 Molecular geometries of model compounds for (A) 1,4 units and (B) 1,2 units. The bond length is shown in Å.

taken as 1.0. We define the dihedral angles (χ) as shown, for example, in Fig. 3.16, that is, the angle between the planes formed by the C_4, C_7, C_4' atoms and the C_7, C_4', C_7' atoms in the 1,4–1,4 dimer and the angle between the planes formed by the C_7, C_6, C_6' and C_6, C_6', C_7' atoms in the 1,2-tail–1,2-tail dimer.

There are 12 possible dimers of 1,2 and 1,4 units, as illustrated in Fig. 3.17. Each 1,2 and 1,4 combination has *cis* and *trans* isomers. The designations 1,2 head and 1,2 tail indicate sequential structures bonded at the 6 position and the 7 position of the 1,2 unit, respectively. The potential energy calculation was carried out for all cases illustrated in Fig. 3.18. Representative results are shown in Fig. 3.18. Figure 3.18(a) gives the conformational potential energy curve for the trans-1,4–1,4 dimer. This combination exhibits two minima at dihedral angles χ_1 of 190° and 260°. Since both potential barriers are very high (85 and 156 kJ/mol), the rotation about the trans-1,4–1,4 bond should be strongly hindered. The difference between the two minima is about 71 kJ/mole. The high potential barriers arise from the steric hindrance of the propyl ring around the main chain. In the *cis*-1,4–1,4 type, only one stable position exists over the rotational angle range. The variations of potential energy with dihedral angle for the

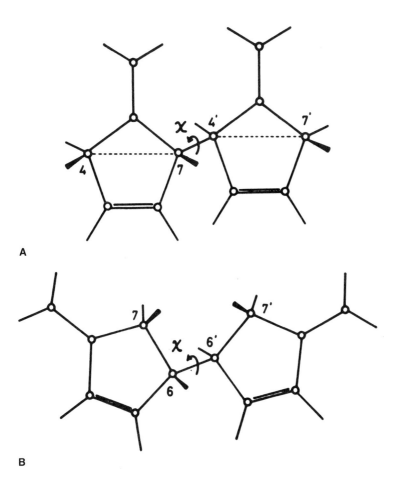

Figure 3.16 The dihedral angle for (A) trans-1,4–1,4 and (B) trans-1,2-tail–1,2-tail isomers.

trans-1,4–1,2 tail and the trans-1,2-tail–1,2-tail types are shown in Fig. 3.18(b) (c), respectively. Since the potential barrier between two minima for the trans-1,2-tail–1,2-tail isomer is much lower than the barriers for the other two cases and the two potential minima are at roughly the same energy, rotational motion between the two potential minima may be possible around the main chain bond. In fact, the barrier in this case is the smallest found among all the dimer species. Thus, a sequence of 1,4 units must be rigid, and the 1,2 unit plays the role of a flexible joint between rigid sequences, with respect to the thermal molecular motion of the chain.

2.2. Aggregation State of PSHD Chains

Figure 3.19 exhibits the temperature dependence of the loss tangent tan δ for the samples with various values of $F_{1,4}$ in a temperature range from 450 to 620 K. The main relaxation process, located at

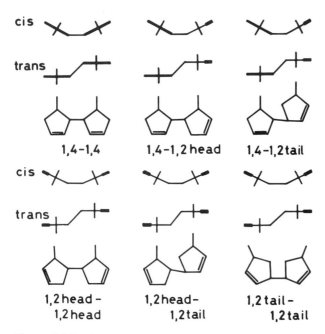

Figure 3.17 Schematic representation of sequential 1,2 and 1,4 units for potential energy calculations.

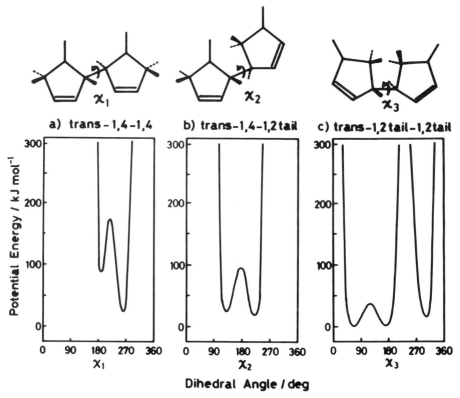

Figure 3.18 Potential energy curves for representative combinations of the 1,2 and 1,4 units.

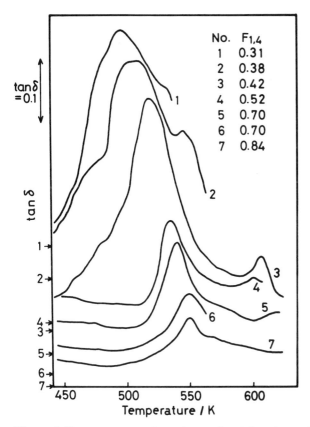

Figure 3.19 Temperature dependence of tanδ for PSHD with various copolymer compositions at 110 Nz. The numbers on the ordinate indicate the zero point of tan δ for the corresponding polymers.

490 K for $F_{1,4} = 0.31$ to 550 K for $F_{1,4} = 0.84$, is called the α process. The storage modulus E' hardly varies with $F_{1,4}$. The location and the magnitude of the α peaks in tan δ depend on the copolymer composition. The α peak shifts to the higher temperature with increasing fraction of 1,4 units. A linear relation is found between the tan δ peak temperature $T_{\tan \delta, \max}$ and $F_{1,4}$, and it is represented by[36,42]

$$T_{\tan \delta, \max} = 470 + 103\, F_{1,4} \tag{7}$$

in K. Figure 3.20 is a plot $F_{1,4}$ versus the α relaxation magnitude evaluated using the equation

$$\Delta E' = \frac{2}{\pi} \frac{E_a}{R} \int_{T_1}^{T_2} E''_{\text{corr}}\, d\left(\frac{1}{T}\right), \tag{8}$$

where E''_{corr} is the value of the dynamic loss modulus, E'' corrected for the background loss, E_a the apparent activation energy for the relaxation, and R the gas constant. The activation energy E_a was assumed to be invariant

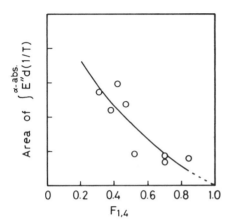

Figure 3.20 Relaxation magnitude for the main absorption as a function of the fraction of 1,4 units.

with $F_{1,4}$; $\Delta E'$ decreases monotonically with increasing $F_{1,4}$ and extrapolates to zero at $F_{1,4} = 1.0$. This trend indicates that the thermal-molecular motion associated with the α mechanical loss is contributed mainly by 1,2-unit sequences, and also that the molecular chain of the 100% 1,4 configuration is incapable of such thermal-molecular motions. From the dynamic viscoelastic behavior and conformational potential energy calculations, it may be concluded that PSHD molecules are composed of rigid segments of the 1,4 units connected by the partially flexible bonds of the 1,2 units.

The density of PSHD decreases with increasing $F_{1,4}$, the fraction of rigid segments. This relation, determined by the least-squares method, is given by

$$\rho = 1.063 - 0.034\, F_{1,4} \quad \text{g/cm}^3. \tag{9}$$

This means that the packing of PSHD molecules becomes looser with increasing fraction of rigid segments. It is reasonable that an increase in length of rigid sequences coupled with a decrease in the number of flexible joints should make chain packing more difficult. Therefore, it can be expected that the packing density of PSHD would be smaller than those of ordinary flexible polymers. Slonimskii, Askadskii, and Kitaigorodskii reported that a linear relation holds between the density and occupied volume of the repeat unit for 70 polymers with widely different chemical structures[43]:

$$\rho = K(M_0/N_A \sum_i \Delta V_i) \quad \text{g/cm}^3, \tag{10}$$

where M_0 is the molecular weight of the repeat unit, N_A is Avogadro's number, and ΔV_i (Å^3) is the volume increment of the ith atomic group in the repeat unit. The value of ΔV_i was calculated from molecular geometry. K

is the packing coefficient, assuming a linear relation between ρ and $(M_0/N_A \sum_i \Delta V_i)$ for all the polymers investigated. The intrinsic volume of the PSHD repeat unit can be obtained as the sum of the volume increments of two CH_2 groups, one aliphatic carbon atom, two aliphatic CH groups, and two CH groups joined by the double bond. The volume increments of the three-membered ring are not listed by Slonimskii et al. Therefore, we calculated the volume increments of two CH_2 groups, and the carbon atom in the ring on the basis of the C–C–C bond angle and the bond length from Slonimskii et al. in order to compare with their equation. The results of calculation are shown in Table 3.2. Figure 3.21 is a plot of ρ against $M/\sum_i \Delta V_i$ as reported by Slonimskii et al., including the data for PSHD (filled circle). The density of PSHD is much lower than would be expected from the general empirical relation. According to Eq. (10), the packing coefficient of PSHD with $F_{1,4} = 0.85$ is 0.646. The average packing coefficient for 70 polymers is 0.681. This unusually low value may be attributed

Table 3.2 Bond Angles and Volume Increments of Atomic Groups

Atom or Atomic Group		Volume Increment (Å³)
CH_2 group of three-membered ring		18.4
Aliphatic carbon atom of three-membered ring		6.34
CH group (aliphatic)		11.1
CH group joined to carbon atom		14.9

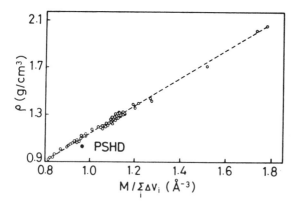

Figure 3.21 A plot of density ρ against $M/\Sigma_i \Delta V_i$. The data for PSHD are shown by the closed circle.

to the large fraction of free (unfilled) space being induced when rigid sequences of 1,4 units are connected by partially flexible 1,2 units. The state of aggregation of PSHD molecules can reasonably explain the density decrease with increasing $F_{1,4}$. However, it may be expected that the magnitude of the density of PSHD increases markedly in the range close to $F_{1,4} = 1.0$ if the 1,2 unit allows the formation of a liquid crystal of 1,4 unit sequences.

The density data lead us to expect that the fraction of free space could be controlled by varying $F_{1,4}$. Perhaps a molecular design for characteristics of the aggregation state and the amount of free space in a solid polymer could be achieved by an ingenious combination of rigid and flexible sequences. It might be interesting to explore the possible application of thin PSHD films as a membrane for selective permeation of gas or liquid.

Figure 3.22 shows x-ray diffractograms of PSHD membranes as a function of $F_{1,4}$. No special orientation of molecular chains was detected in the directions normal and parallel to the membrane surface, indicating completely random-chain orientation. A broad intensity distribution of amorphous halo having multiple peaks was observed for the samples with lower $F_{1,4}$ ($= 0.30$ and 0.43). For the samples with higher $F_{1,4}$, the amorphous halo was sharper. These facts indicate that the samples with lower fractions of 1,4 units have a broad distribution of mean intermolecular distances. It is apparent that the molecular aggregation state of PSHD is heterogeneous, that is, consisting of two regions in which the chains are densely and loosely packed. The aggregation state is more homogeneous for samples with higher $F_{1,4}$.

Figures 3.23 and 3.24 show electron micrographs of surface replicas of PSHD membranes with $F_{1,4}$ values of 0.30 and 0.75, respectively. The ion-etching technique was employed to investigate the heterogeneity of molecular chain packing.[44] The etching time was varied to clarify the effect of ion etching. The surface of the specimen became clearer by increasing the

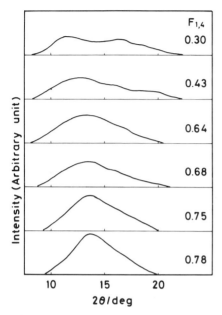

Figure 3.22 X-ray diffractograms of PSHD with various $F_{1,4}$ values.

degree of ion etching. A remarkable difference in the morphology of the etched surface was observed between samples with higher and lower fractions of 1,4 units. Though the sample with $F_{1,4} = 0.30$ initially had a relatively smooth surface, uneven surfaces composed of granular aggregates of diameter about 20–40 nm became apparent with increasing etching time. These granular structures were similar to those reported by Frank et al. for polycarbonate specimens ion etched after annealing below the glass-transition temperature.[44] They pointed out that these grains are composed of regions characterized by a certain parallelism of chains. In the samples with low fractions of 1,4 units, however, the PSHD chains may not align in parallel but aggregate in a manner similar to random coils, since the bond angle of the 1,2 unit joining to 1,4-unit sequences is acute, and this situation induces globular aggregates of chains during the chain-propagation reaction. It is reasonable to consider that loosely packed regions are less stable for ion etching than densely packed ones, the former being destroyed by ionic collision even during a brief ion-etching treatment. Therefore, these granular structures must correspond to domains in which molecular chains are densely packed, but their conformations or intermolecular orders are not yet clear. The surface morphology of the ion-etched samples may indicate the degree of heterogeneity of molecular chain packing, although it does not necessarily represent the internal structure. For the sample with $F_{1,4} = 0.75$, granules of about 10 nm in diameter were observed after etching for 15 min, and the surface became gradually

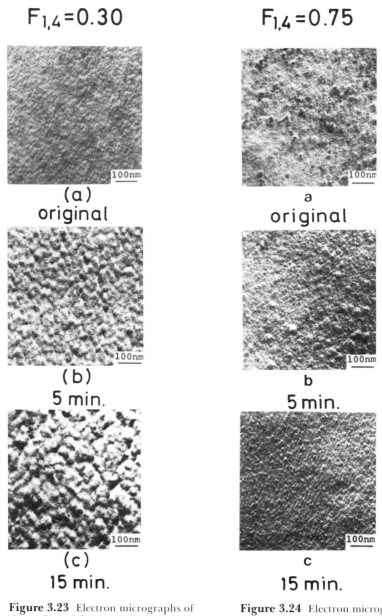

Figure 3.23 Electron micrographs of the surface of the PSHD films with $F_{1,4}$ = 0.30 that were ion etched for (a) 0, (b) 5, and (c) 15 min.

Figure 3.24 Electron micrographs of the surface of the PSHD films with $F_{1,4}$ = 0.75 that were ion etched for (a) 0, (b) 5, and (c) 15 min.

smoother as the ion etching advanced. These observed morphologies of the surface replicas indicate that the heterogeneity with respect to molecular aggregation depends on the fraction of the rigid sequences. The state of chain aggregation in a PSHD sample with $F_{1,4} = 0.30$ was more heterogeneous than that in the $F_{1,4} = 0.75$ sample. This agrees well with the x-ray diffraction study shown in Fig. 3.22.

The heterogeneity in the state of molecular aggregation was also confirmed by dynamic viscoelastic measurements.[36] The α relaxation process, which is associated with thermal motion of the sequential portion connected by a 1,2 unit, was observed in tan δ in a temperature range from 450 to 620 K. The multiple α peaks in tan δ were clearly observed for samples with lower fractions of 1,4 units, and they gradually changed into a single peak by increasing the fraction of 1,4 units. A single relaxation process was observed for samples with $F_{1,4}$ more than 0.52 in a temperature range from 530 to 550 K. The appearance of the multiple α peaks is attributed to the heterogeneity of molecular aggregation.

Figure 3.25(a) and (b) shows schematically the principal features of aggregation states of PSHD chains. Figure 3.25(c) shows a random-coil model of flexible chains. The bold and broken lines correspond to the rigid and semirigid sequences, respectively. Figure 3.25(a) shows the state of microscopically homogeneous aggregation (polymer chain with long rigid sequences), while Fig. 3.25(b) shows the state of microscopically heterogeneous aggregation (polymer chains with short rigid sequences). When long rigid sequences are connected by semirigid ones, as shown in Fig. 3.25(a), the free space increases due to the extremely restricted conformations of the chain. In the extreme case of $F_{1,4} = 1.0$, the density would appreciably increase due to possible parallel alignment of the 1,4-unit sequences, which may form a liquid crystal. For PSHDs with smaller $F_{1,4}$, the PSHD chains contract themselves, and globular aggregation is more apparent, as sche-

Figure 3.25 Models of molecular chain aggregation for (a) polymers with long rigid sequences connected by semirigid sequences, (b) polymers with short rigid sequences connected by semirigid sequences (—— rigid sequence, --- semirigid sequence), and (c) flexible polymers (random coil).

matically shown by the change from Fig. 3.25(a) to (b). The propagating ion pair of the molecular chain may be confined in a globular molecular coil owing to the conformational restriction. This is suggested from the fact that the number-average molecular weight decreases with decreasing $F_{1,4}$. The radius of gyration of the chains composed of 1,2 units may contract due to the globular characteristics resulting from the acute bonding angle and the restricted rotational angles of the 1,2 units. Therefore, the chain expansion in the sample with $F_{1,4} = 0.30$ is probably comparable to or less than that of unperturbed chains of flexible polymers. The morphological observation of Fig. 3.23(c) apparently indicates that there exists microheterogeneity composed of the grain region of high density and the intergrain region of low density. Therefore, we propose the model of Fig. 3.25(b), in which a globular molecular coil entangles itself to form a dense core and also, a loose shell interpenetrates the loose shells of the other molecules.

2.3. Relationships of Molecular Packing to Gas Permeation in PSHD

The permeability of gases through PSHD membranes increases with increasing temperature. This indicates that the PSHD membrane may be treated as a uniform membrane. We assume that Fick's law is obeyed in this system, because the interaction between the polymer and gas molecules studied here may be small. The relationship $P = DS$ may be applied to PSHD membranes, where P is the permeability coefficient, D the diffusion coefficient, and S the solubility coefficient. Of course, in order to clarify the permeation mechanism of gases in the PSHD membrane, P, D, and S should be evaluated separately.[46]

Table 3.3 shows the permeability coefficients of H_2, He, O_2, Ar, and N_2 gases, the diffusion coefficient and the solubility coefficient of N_2 gas, and the ratio of the permeability coefficient of each gas to that of N_2 gas (P/P_{N_2}) at 298 K as a function of $F_{1,4}$.

Figure 3.26 shows the $F_{1,4}$ dependence of the permeability coefficients for various gases at 298 K. The characteristic trend is the appearance of a minimum in the vicinity of $F_{1,4} = 0.7$. A similar trend was also observed for the diffusivity of N_2 gas at 298 K, as shown in Fig. 3.27. Figure 3.28 shows a semilogarithmic plot of the solubility coefficient for N_2 gas against $F_{1,4}$ at 298 K. The solubility coefficients do not vary with $F_{1,4}$. The results imply that the $F_{1,4}$ dependence of the permeability coefficient mainly arises from that of the diffusion coefficient. Although the chain-conformation–solubility relationship is not clear at the moment, the conformational difference of 1,2 and 1,4 units must have a direct effect on the diffusion process but not on the solution characteristics.

The diffusion process is affected by the following factors: (1) the packing state of molecular chains and the heterogeneity of molecular aggregation and (2) the degree of thermal-molecular motions. Because an increase in

Table 3.3 Characteristics of Gas Permeation in PSHD Membranes

No.	$F_{1,4}$	$P_{H_2} \times 10^9$	$P_{He} \times 10^9$	$P_{O_2} \times 10^{10}$	$P_{Ar} \times 10^{10}$	$P_{N_2} \times 10^{11}$	$D_{N_2} \times 10^8$	$S_{N_2} \times 10^3$	P_{H_2}/P_{N_2}	P_{He}/P_{N_2}	P_{O_2}/P_{N_2}	P_{Ar}/P_{N_2}
		$\left(\dfrac{cm^3\,(STP)\,cm}{cm\,s\,(cm\,Hg)}\right)$					$\left(\dfrac{cm^2}{s}\right)$	$\left(\dfrac{cm^3\,(STP)}{cm^3\,(cm\,Hg)}\right)$				
1	0.78	3.73	2.86	4.51	2.67	8.13	2.26	3.60	45.9	35.2	5.55	3.28
2	0.75	2.97	2.42	3.60		6.81	1.60	4.26	43.6	35.5	5.29	
3	0.73	2.96	2.40	2.85	1.71	6.18	1.52	4.07	47.9	38.8	4.61	2.77
4	0.68			3.53		6.37					5.54	
5	0.64	2.77	2.09	3.67		6.44	1.65	3.90	43.0	32.5	5.70	
6	0.46		2.52			10.7	3.74	2.86		25.0		
7	0.30	5.82	4.81	6.33	3.47	19.1	4.43	4.31	30.5	25.1	3.31	1.82

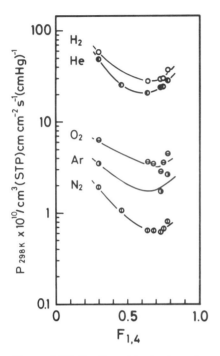

Figure 3.26 Semilogarithmic plots of permeability coefficients for H_2, He, O_2, Ar, and N_2 gases versus $F_{1,4}$ at 298 K.

free space within the membrane enhances the gas diffusivity, a decrease in the density with increasing $F_{1,4}$ should give rise to a monotonic increase of D. However, the $F_{1,4}$ dependence of D is the reverse of this expectation for the samples with smaller $F_{1,4}$, as shown in Fig. 3.27. Therefore, we must find another reason to explain the unexpected $F_{1,4}$ dependence of the gas diffusivity. This behavior cannot be explained by a variation in thermal-

Figure 3.27 Semilogarithmic plots of diffusion coefficients for N_2 gas versus $F_{1,4}$ at 298 K.

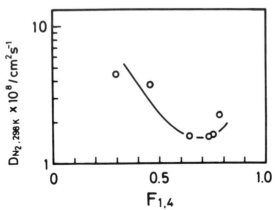

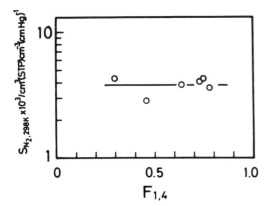

Figure 3.28 Semilogarithmic plot of solubility coefficients (P/D) for N_2 gas versus $F_{1,4}$ at 298 K.

molecular motions either. The tan δ peak temperature for the α relaxation process of PSHD studied here is much higher than the measuring temperature 298 K, at which the thermal motion is almost frozen. Therefore, the variation in thermal motion does not so much affect the diffusivity of gas as to explain the $F_{1,4}$ dependence of D. We consider the heterogeneity of molecular aggregation to elucidate this dependence. It is apparent that when two specimens with the same density are compared, the diffusivity in the one with the more heterogeneous matrix is greater than that in the other with a less heterogeneous matrix, since gas molecules prefer to permeate through more loosely packed regions. The packing of PSHD molecular chains changes from heterogeneous to homogeneous with increasing $F_{1,4}$, as discussed above. Thus the variations in density and heterogeneity with $F_{1,4}$ may lead to a minimum of the $F_{1,4}$ dependence on the diffusion or the permeability coefficient. As mentioned above, the diffusion of gas through the polymeric membrane sensitivity reflects the internal structure of the chain packing or its heterogeneity as well as thermal-molecular motion of polymeric chains. Therefore, it is worthwhile to examine the internal structure within polymeric membranes in terms of diffusion and/or permeation properties, in addition to x-ray studies or morphological observations.

Figure 3.29 shows semilogarithmic plots for the permeability coefficients for H_2, He, Ar, and O_2 gases relative to that of N_2 gas against $F_{1,4}$ at 298 K. Logarithms of P_{H_2}/P_{N_2}, P_{He}/P_{N_2}, P_{Ar}/P_{N_2}, and P_{O_2}/P_{N_2} vary linearly with $F_{1,4}$. It is noteworthy that these ratios exhibit no maximum at the minimum point for P in Fig. 3.26 (corresponding to the arrows in Fig. 3.29) nor decrease with increasing P_{N_2} in the range of higher $F_{1,4}$. It has been reported that the ratios of P for several gases to P_{N_2} decrease with increasing P_{N_2} for membranes of butadiene–acrylonitrile copolymer and for membranes of

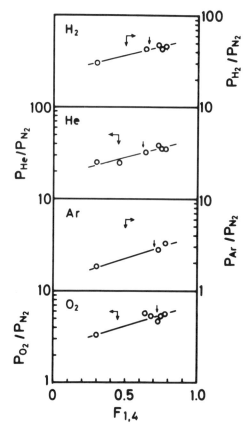

Figure 3.29 Semilogarithmic plots of the ratios of permeability coefficients for various gases to nitrogen gas versus $F_{1,4}$ at 298 K.

silicone and polycarbonate blend. Such a trend is generally recognized. However, for PSHD membranes with high $F_{1,4}$, both permeabilities and permselectivities of gases increase with increasing $F_{1,4}$.

Figure 3.30 shows plots of P/P_{N_2} against P_{N_2} at 298 K for various polymers quoted from the *Polymer Handbook*.[2] The data for the polymers in the rubbery state exhibit an inverse proportionality relationship on the double logarithmic plots, as shown by open circles, except for vulcanized poly(oxydimethylsilylene) containing 10 wt % filler. On the other hand, the data for the polymers in the glassy state are scattered, as shown by half-filled circles. It is often found that the activation energy for permeation or diffusion changes discontinuously in the temperature range of a glass transition.[1-3] This implies that the mechanism of diffusion or permeation distinctly changes with variations in thermal-molecular motion. These two types of experimental results indicate that there is a distinct difference in

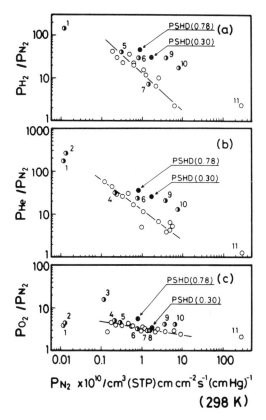

Figure 3.30 A plot of the permeability coefficient ratio of (a) hydrogen, (b) helium, and (c) oxygen to nitrogen with permeability coefficient of nitrogen at 298 K for (○) rubbery polymers, (●) glassy polymers, and (1) vulcanized poly (oxidimethylsilylene) with 10 wt % filler.

the permeation mechanism of gases between glassy and rubbery polymers. The inverse proportionality of permselectivity versus permeability shown in Fig. 3.30 indicates that the remarkably large permeability and permselectivity values cannot be expected for homogeneous polymeric membranes in the rubbery state. Since the permselectivity of glassy polymers deviates from this relation, superior permeability and permselectivity characteristics should be obtained with polymeric membranes in the glassy state rather than in the rubbery state. The ratios of the permeability coefficients of H_2, He, and O_2 to N_2 gases for the PSHD membranes of $F_{1,4} = 0.30$ and 0.78 are also plotted in Fig. 3.30. Since these ratios are considerably higher than those of other polymeric membranes except for P_{O_2}/P_{N_2} for $F_{1,4} = 0.30$, a glassy polymer containing rigid sequences may be utilized as a permselective membrane with excellent heat resistance.

3. Phase Transition and Molecular Orientation Effects on Gas Separation through Polymer/LC Composite Films

The effect of thermal-molecular motion on permeation characteristics is an interesting problem from the standpoint of the diffusion behavior of permeable molecules. The relationship between permeabilities of water or gases and thermal-molecular motion of polymeric chains has been extensively reported.[4,34,35] It was found that water or gas permeations abruptly increased at a temperature range corresponding to the primary relaxation process or the phase transition of the membrane polymer. Biological membranes are composed of various types of phospholipids, cholesterols, and functional proteins. Bimolecular membranes of lipids are in a liquid-crystalline state, being capable of reversible structural modifications, and the permeation properties of biomembranes depend upon such reversible changes.[47,48] Synthetic polymeric membranes exhibit permeation characteristics to gases and liquids depending on the sorption mechanism of permeants on the membrane surface and the diffusion mechanism of holes, in other words, free volume.

The purpose of this section is to investigate the structural properties or compatibilities of the composite membrane composed of polymer and liquid-crystalline material, the relationship of thermal-molecular motion to permeation, and also the permeation mechanism of the composite membrane to hydrocarbon gases or cationic ions. Electric field effects on permeation of isomeric hydrocarbon gases have also been studied.

3.1. Construction and Aggregation State of Polymer/LC Composites

Polymer/liquid-crystal composite membranes were cast from a 1,2-dichloroethane solution of polymer and liquid-crystalline materials shown in Fig. 3.31. Polycarbonate (PC) and poly(vinyl chloride) were used as the matrix substance for the composite membrane because it does not exhibit any apparent thermal-molecular motion or phase transition in the temperature range studied here. The polymer/liquid-crystal composite membranes contain liquid crystal of 15, 30, 45, and 60 wt %. These composite membranes are designated the 85/15, 70/30, 55/45, and 40/60 composite membranes, respectively. The composite membranes were cured at 333 K for 2 h in vacuo prior to gas permeation experiments.[4,34,35]

In order to elucidate an aggregation or dispersing state of EBBA molecules in a composite membrane, thermal analyses are very useful. Figure 3.32 shows the differential scanning calorimetric (DSC) curves for PVC, EBBA, and PVC/EBBA composite membranes. Two endothermic peaks were observed for EBBA, which correspond to the crystal–nematic (T_{KN} = 304 K) and nematic–isotropic (T_{NI} = 355 K) transition temperatures, re-

(a) POLYMER
  1) Polycarbonate (PC)

  2) Poly(vinyl chloride) (PVC)

(b) LIQUID CRYSTAL
  1) N-(4-ethoxybenzylidene)-4'-butylanilline (EBBA)

  $CH_3CH_2O\text{-}\bigcirc\text{-}CH=N\text{-}\bigcirc\text{-}(CH_2)_3CH_3$

  K→N 304 K, N→I 355 K

  2) 4-cyano-4'-pentyl biphenyl (CPB)

  $C_5H_{11}\text{-}\bigcirc\text{-}\bigcirc\text{-}C\equiv N$

  K→N 296 K, N→I 308 K

  3) 4-cyano-4'-heptyloxy biphenyl (CHOB)

  $C_7H_{15}O\text{-}\bigcirc\text{-}\bigcirc\text{-}C\equiv N$

  K→N 325 K, N→I 344 K, $\Delta\varepsilon = +12$ (323 K)

  4) 2-Sub.-1, 4-bis-(4-n-pentylbenzoyloxy)-benzene (X-BPBB)

  X = H Me Cl Br t-Bu
  T_{KN}/K 398 353 354 344 (316)
  T_{NI}/K 459 416 415 387 (328)

(c) FLUOROCARBON
  1) Perfluorotributylamine (PFTA) $(CF_3CF_2CF_2CF_2)_3N$
  2) Tris(1 H, 1 H, 5 H-octafluoropentyl) phosphate (TPP)
     $(HCF_2CF_2CF_2CF_2CH_2O)_3PO$

(d) AZOBENZENE-LINKED CROWN ETHER
  1) AZO-CR(1)

  2) AZO-Cr(2)

(e) AMPHIPHILIC CROWN ETHER
  1) AMP-CR(a), AMP-CR(b)

  a, $R = CH_3(CH_2)_{10}\text{-}$
  b, $R = (CH_3(CH_2)_{15}OCH_2)_2CHOCH_2\text{-}$

  2) AMP-CR(c)

  3) AMP-CR(d)

  $R = Me(CH_2)_2CO$

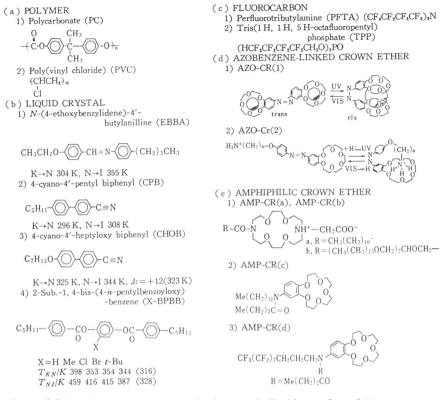

Figure 3.31 (a) Chemical structures of polymers, (b) liquid crystals, and (c) fluorocarbon monomers.

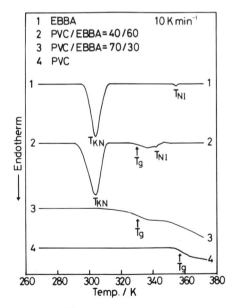

Figure 3.32 DSC curves of PVC (4), EBBA (1), and PVC/EBBA [(2) 40/60; (3) 70/30 composite membranes. T_g represents the glass transition temperature of PVC.

spectively (curve 1 in Fig. 3.32). These transition processes were also observed at approximately similar temperature ranges for the PVC/EBBA composite membranes at which the EBBA fraction was 45–70 wt % [curve 2 for the 40/60 (PVC/EBBA) composite membrane]. However, the PVC/EBBA composite membranes in which the EBBA fraction was below 30 wt % did not exhibit such transition peaks on the DSC curve, but this curve deviated from a flat baseline at about 300 K (curve 3). This thermal behavior indicates that EBBA is molecularly dispersed below 30 wt %. Therefore, in the case of the EBBA fraction above 45 wt %, it is clear that EBBA exists in both states of molecular dispersion and crystal domains. The glass transition temperature T_g for PVC homopolymer was observed at about 360 K (curve 4), and it was depressed to about 330 K for the composite membranes (curve 2 and 3). The lowering of T_g indicates that PVC and EBBA are quite miscible and EBBA in the composite membranes acts as a plasticizer for PVC.

Figure 3.33 shows the scanning electron microscopic (SEM) photographs of the fracture surface (upper portion in each photograph) and membrane (lower portion) for the 40/60 (PVC/EBBA) composite membrane quenched with liquid N_2 after annealing at 333 K ($> T_{KN}$). Annealing was performed in order to observe the surface states of the composite membrane above T_{KN}, since the gas-permeation property is markedly affected by the surface states of the membrane, especially above T_{KN}. AFS and SFS represent the air-facing and the substrate-facing surfaces when the membrane was cast, respectively. It is apparently observed in Fig. 3.33(a) that EBBA domains

Figure 3.33 Scanning electron micrographs of 60 wt % EBBA composite membrane quenched at a temperature in a liquid-crystalline state (a) and fracture surface (FS) (upper portion in each photograph) and membrane surface (lower portion) for the 40/60 (PVC/EBBA) composite membrane after extracting EBBA with ethanol at 333 K for 2 h (b). AFS and SFS represent air-facing and substrate-facing surface, respectively.

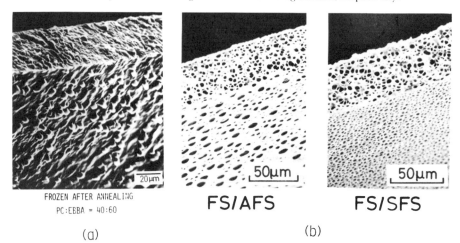

are oozing from the membrane surface and spreading over the whole area. The sample after extracting EBBA with ethanol at 333 K for 2 h presents the appearance of a spongy material, as shown in Fig. 3.33(b) and (c). Since about 95 wt % of total EBBA was extracted from the composite membrane with hot ethanol, it is evident that EBBA molecules form domains as an interpenetrating continuous phase among the three-dimensional spongy networks of PVC matrix. Consequently, it seems reasonable to consider that the continuous EBBA phase takes the role of a main diffusing region for gas permeation. In addition, the concentration of EBBA on AFS was relatively a little higher than that on SFS. Thus, the gas-permeation experiments in this study were consistently performed by use of the AFS side as a gas feed surface.

Furthermore, the formation process of a composite membrane was traced with a polarizing optical microscope (POM). Figure 3.34 shows the POM photographs for the 40/60 (PVC/EBBA) composite membrane at a temperature of 293 K ($< T_{KN}$) upon solvent evaporation. In the case of an isotropic liquid state just after casting, a field of vision was dark [Fig. 3.34 (1)]. Liquid-crystalline molecules gradually clustered or aggregated with solvent evaporation, resulting in an exclusion of PVC molecules, as shown by a dark field in Fig. 3.34 (2). At a certain concentration, the crystallitelike domains of EBBA were produced and began to grow in PVC matrix [Fig. 3.34 (3)–(8)]. Simultaneously, an excluded PVC phase surrounded EBBA crystal domains as the networks in a similar manner to a coacervation effect.[49,50] Consequently, distinct crystal domains of EBBA were formed in the composite membrane as a continuous phase, and the characteristic in-

Figure 3.34 Polarizing optical micrographs under crossed nicols during solvent evaporation for the 40/60 (PVC/EBBA) composite membrane at 293 K ($<T_{KN}$).

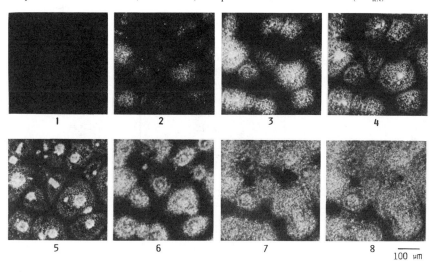

terpenetrated structure composed of both components was completed in this process. Of course, the diameter of the EBBA cluster in Fig. 3.34 is greater than that of the holes in Fig. 3.33. The EBBA clusters may be divided into smaller ones, although the surrounding PVC networks cannot be distinguished under POM.[35]

Figure 3.35 shows stress–strain curves for a series of PVC/EBBA composite membranes at several weight fractions of EBBA (at 293 K, $< T_{KN}$). The initial modulus E for the PVC/EBBA composite membranes decreased with an increase in the weight fraction of EBBA, but the strain at the break, ε_b, increased up to 45 wt % of EBBA in comparison with that for PVC homopolymer (100/0). Also, the values of E for the composite membranes above the EBBA fraction of 60 wt % increased again, which might be due to the contribution from the continuous crystal phase of EBBA. In particular, the 40/60 (PVC/EBBA) composite membrane, for which the gas-permeation characteristics were investigated in this study, exhibited an excellent ductility; that is, $\varepsilon_b = 413\%$. In the case of practical applications as a permselective membrane, the ductile properties of the membrane are required from a viewpoint of preparation of ultrathin films. Then the mechanical properties for the 40/60 (PVC/EBBA) composite membrane proved to suffice for this requirement. As mentioned, the composite membrane is composed of the interpenetrating continuous phases of liquid-crystalline molecules and polymer matrix. This kind of aggregation state should provide excellent mechanical stability properties to composite membranes.

Figure 3.36 shows the sorption isotherms of CH_4, C_3H_8, and C_4H_{10} for the 40/60 composite membrane. The measurements were carried out above

Figure 3.35 Stress–strain curves for a series of the PVC/EBBA composite membranes at several weight fractions of EBBA (at 293 K, $<T_{KN}$):(1) 100/0; (2) 85/15; (3) 70/30; (4) 55/45; (5) 40/60; (6) 30/70.

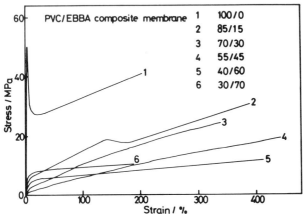

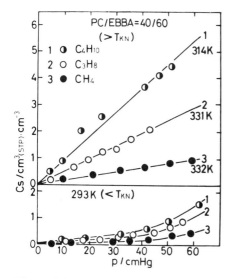

Figure 3.36 Sorption isotherms of hydrocarbon gases for the 60 wt % EBBA composite membrane below and above the phase transition temperature T_{KN}. (1) (◐) C_4H_{10}, (2) (○) C_3H_8, (3) (●) CH_4.

and below the crystal–nematic phase transition temperature of EBBA, T_{KN}. The validity of Henry's law for the sorption isotherm of these hydrocarbon gases below T_{KN} is doubtful, but above T_{KN} it obeys the law. Calculated values of the solubility coefficients for CH_4, C_3H_8, and C_4H_{10} in the composite membrane are 1.69×10^{-2}, 4.62×10^{-2}, and 1.26×10^{-1} cm³ (STP/cm³ cm Hg), respectively. Therefore, the solubility of C_4H_{10} is about 7.5 times greater than that of CH_4.

Figure 3.37 shows sorption–desorption curves for CH_4 in the 40/60 composite membrane. Here, $M(t)$ is the total quantity of penetrant that has entered or left the membrane in time t, and $M(\infty)$ is the corresponding quantity at infinite time, that is, when equilibrium is reached between sorption and desorption. The linear relationship between $M(t)/M(\infty)$ and $t^{1/2}$ was recognized in the early stage of an experiment at 330 K (above T_{KN}), and this shape of the experimental curve suggests that this system follows Fickian sorption.[51] Therefore, in the temperature range above T_{KN}, it is reasonable to consider that a steady surface equilibrium is established immediately and that the diffusion coefficient of CH_4 is a function of the concentration only. Similar behavior is observed for the polymer–organic-vapor system in the rubbery state. In addition to the results of x-ray, DSC, and sorption measurements mentioned here, these sorption or desorption experiments apparently indicate that the 40/60 composite membrane can be handled as a homogeneous medium when considering gas permeation.[5,29,52]

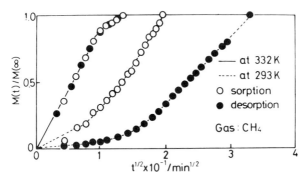

Figure 3.37 (○) Sorption– (●) desorption curves of CH_4 for the 60 wt % EBBA composite membrane et (---) 293 and (——) 332 K (gas: CH_4).

3.2. Phase Transition Effects on Gas Permeation through Polymer/LC Composite Membranes

Figure 3.38 shows an Arrhenius plot of the permeability coefficients P of He, N_2, and various hydrocarbon gases for the 40/60 composite membrane. A distinct jump in P was observed in the vicinity of the transition temperature of EBBA from the crystalline to the nematic phase. In particular, the magnitude of P for hydrocarbon gases increases by approximately 100–500 times over a few degrees in the phase transition region, shown by the broken lines in Figure 3.38. In this region, the permeabilities of He and N_2 gases also increase markedly. As He gas is inert, its solubility properties in the composite membrane may not change over the temperature range of the phase transition. Therefore, this discontinuous increase of P for an inert gas may be caused by a marked increase in the diffusion coefficient. This tendency may arise from the activation of thermal-molecular motion of the membrane and/or an increase in hole formation. The probability for hole formation is enhanced due to a dynamic phase equilibrium within the EBBA domain in going from the crystal to the liquid-crystalline states or vice versa. The dynamic phase equilibrium means coexistence of solid and liquid-crystalline phases with respect to time and position during the process of the phase transition.

With respect to permeation of various hydrocarbon gases, the magnitudes of P increase in the order of i-C_4H_{10}, C_4H_{10}, C_3H_8, and CH_4 in the temperature range below the phase transition temperature of EBBA and in the order of CH_4, C_3H_8, i-C_4H_{10}, and C_4H_{10} in the temperature range above the transition. The variations of P with temperature for CH_4, C_3H_8, and C_4H_{10} are nearly reversible across the phase transition range. In other words, below the phase transition temperature, the magnitudes of P decrease with increasing number of carbon atoms or the occupied volume of gas, and this trend may suggest that the permeation process is predominantly governed by the diffusion process. On the other hand, above the

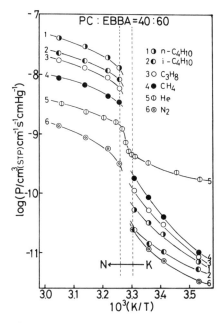

Figure 3.38 Arrhenius plot of P for He, N_2, and hydrocarbon gases for the PC/EBBA (40/60) composite membrane. (1) (◐) n-C_4H_{10}, (2) (◐) i-C_4H_{10}, (3) (○) C_3H_8, (4) (●) CH_4, (5) (◐) He, and (6) (○) N_2.

phase transition temperature of EBBA, the magnitudes of P increase with an increasing number of carbon atoms. Generally speaking, this tendency cannot be explained by the diffusion control mechanism, but it is reasonably expected that permeability is governed mainly by the solubility process.

The solubility coefficient S can be evaluated from the permeability P and the diffusion D. Figure 3.39 shows an Arrhenius plot of the solubility coefficients for CH_4, C_3H_8, C_4H_{10}, and i-C_4H_{10} in the 40/60 composite membrane. The magnitudes of S increase with increasing temperature up to the neighborhood of the phase transition temperature exhibited by the broken line, and decrease gradually after the maximum S. The negative and positive slopes of the Arrhenius plot correspond to positive and negative heats of solution for S, respectively, and this behavior of log S versus $1/T$ apparently indicates a variation of the solubility mechanism of penetrant gases to the polymer/liquid-crystal composite membrane across the phase transition region. The solubility mechanism must vary depending on the fractions of polymer and EBBA in the surface layer because hydrocarbon gases dissolve into the polymer and EBBA to different extents. Beyond the phase transition temperature range, however, the magnitudes of S increase by about 6 times for CH_4, 10 times for C_3H_8, and 20 times for C_4H_{10} and i-C_4H_{10}, and also the absolute value of S increases in the order CH_4, C_3H_8,

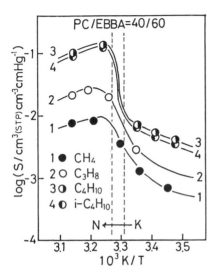

Figure 3.39 Arrhenius plot of S for hydrocarbon gases for the PC/EBBA (40/60) composite membrane. (1) (●) CH_4, (2) (○) C_3H_8, (3) (◐) C_4H_{10}, and (4) (◑) i-C_4H_{10}.

and C_4H_{10} or i-C_4H_{10} in the whole temperature range studied here.[34,55] Inasmuch as the thermal expansion coefficients of polymer and EBBA are quite different, especially in the case of the liquid-crystalline state of EBBA, it is reasonable to consider that the weight fractions of polymer and EBBA, or their mixture state in the surface layer of the membrane, vary with measured temperature. Based on the scanning electron microscopic observation as shown in Fig. 3.33(a), it is apparent that the EBBA domain alternately comes in and out of the membrane surface below and above the phase transition temperature, respectively, due to the larger volume expansion coefficient of EBBA than that of the polymer.

3.3. Molecular Filtration of Hydrocarbon Isomers through Polymer/LC Composite Membranes Based on Molecular Orientation

When hydrocarbon gases diffuse through the membrane, their longer molecular axes align in the diffusing direction. Under such circumstances the diffusion resistance (or friction) to hydrocarbon gases is strongly influenced by the magnitude of the cross-sectional dimension normal to the main axes, as illustrated in Fig. 3.40.[53,54] Since the effective cross-sectional dimensions (diameters) of n-, iso-, and neo-hydrocarbon isomers are 0.49, 0.56, and 0.62 nm, respectively, it is generally difficult to separate hydrocarbon isomers by ordinary synthetic membranes, due to the inability to control the cross-sectional dimensions of the free volume or channel to within several Å.

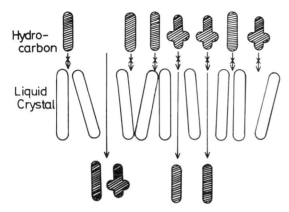

Figure 3.40 Illustration of the permselectivity for isomers with different cross-sectional areas.

The orientation of liquid-crystalline molecules has a direct effect on the permselectivity of hydrocarbon isomers on the basis of the path length for diffusion and the distribution of molecular channel dimension, as illustrated in Fig. 3.41. In the case of a well-oriented state of liquid-crystalline molecules, the permeant gases can diffuse along quite a straight path, resulting in a greater flux. Since hydrocarbon isomers permeate through intermolecular channels in the liquid-crystalline phase for the polymer/LC composite membrane, it is expected that hydrocarbon isomers can be effectively separated through a liquid-crystalline medium in which the intermolecular channel dimensions can be controlled within several Å with a narrow distribution. We have reported with voltage dependence of the permeation of n-C_4H_{10} and iso-C_4H_{10} through the polymer/LC composite membrane.[55,56]

Figure 3.41 Schematic representation of the path for diffusion of permeant molecules in the oriented and unoriented composite membranes.

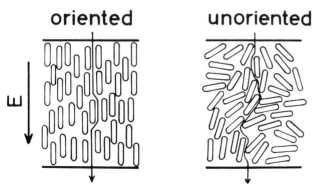

The poly(vinyl chloride)(*PVC*)/4-cyano-4'-pentylbiphenyl (CPB) composite membrane contains CPB of 60 wt %, which has the positive dielectric anisotropy due to the direction of the dipole moment of the $-C \equiv N$ group along with the molecular axis. Then, when the electric field is applied to it, CPB molecules are oriented preferentially in the direction of the electric field. A pair of electrodes was set in the permeation cell, and, therefore, an electric field could be applied during measurements of the gas permeation. The space between the electrodes was 5 mm.

Figure 3.42 is the wide-angle x-ray diffraction photograph taken when an electric field is applied to the composite membrane. The inner and outer reflections correspond to the molecular length of CPB and the intermolecular distance, respectively. When an electric field was applied perpendicular to the membrane surface, it emphasized the normal alignment to the film surface. Since the liquid-crystalline material aggregates in a continuous phase, it may be possible to control the diffusivity of gas molecules through the composite membrane by application of an electric field on the basis of the relationship between the dimensions of the channels formed in the intermolecular regions and the sectional dimensions of the permeating gases.

Figure 3.43 shows the dependence on the applied voltage of the permeability coefficients P for C_4H_{10} and iso-C_4H_{10}. The difference in diameter of the sections of C_4H_{10} and iso-C_4H_{10} is only 0.07 nm ($d_{C_4H_{10}}$ = 0.49 nm and $d_{iso-C_4H_{10}}$ = 0.56 nm). As the applied voltage was increased, the magnitude of P for both C_4H_{10} and iso-C_4H_{10} gradually increased. The effect on P of the applied voltage can be explained by the relative difference of diffusion paths. Gas molecules diffuse along a fairly straight path in an oriented composite membrane and, on the other hand, diffuse along a tortuous path in an unoriented membrane. The separation ratio of $P_{C_4H_{10}}$ to $P_{iso-C_4H_{10}}$, α, was 2 at zero voltage (random orientation of liquid-crystalline molecules), and increased to 5 at 730 V (oriented state) when the space

Figure 3.42 Wide-angle x-ray diffraction of 60 wt % CPB composite membrane. The x-ray pattern was taken under the conditions of an applied electric field and above T_{KN}.

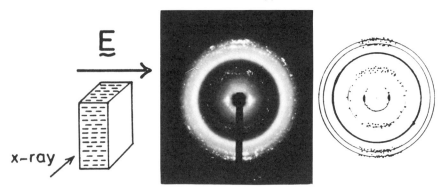

Phase Transition and Molecular Orientation Effects on Gas Separation

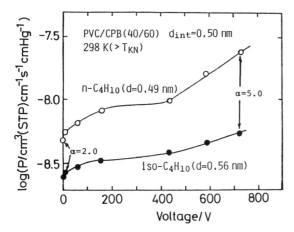

Figure 3.43 Variation of the permeability coefficients for C_4H_{10} and iso-C_4H_{10} with the magnitude of the applied voltage for the PVC/CPB (40/60) composite membrane.

between the two electrodes was 5 mm. The increase of α with the magnitude of the applied voltage indicates that the distribution of channel dimensions among LC molecules becomes narrower as a result of a stronger molecular orientation perpendicular to the membrane surface. This result indicates that the greater degree of orientation of LC molecules induces a more effective separation of C_4H_{10} and iso-C_4H_{10}, even though the difference of their molecular diameters is less than 1 Å.

Figure 3.44 is the wide-angle x-ray diffraction photographs of the PVC/CHOB (30/70) composite membranes prepared without and with an imposed electric field during evaporation of the solvent. When the oriented composite membrane (E = 2350 V/cm) was placed as shown in Fig. 3.44 and an x-ray beam was passed parallel to the membrane surface (edge view), x-ray reflections corresponding to the length of a CHOB molecule or dimer were observed on the equator. The reflection at 2.16 nm corresponds to the length of a CHOB molecule and those at 3.51 and 2.94 nm to the length of the dimer. Figure 3.44 indicates that the CHOB molecular axes preferentially align perpendicular to the membrane surface during evaporation of solvent from a solution under an imposed electric field, in other words, parallel to the direction of an imposed electric field due to a positive dielectric anisotropy. The Fig. 3.44(c) photograph, taken after the permeation experiment under an imposed electric field of 50000 V/cm perpendicular to the surface of the oriented composite membrane, demonstrates that orientation of CHOB molecules in the composite membrane was maintained even after the permeation experiment above T_{KN}.

Figure 3.45 shows an Arrhenius plot of the permeability coefficient P for C_5H_{12} in the oriented and the unoriented PVC/CHOB (30/70) composite membranes and also the permeability coefficient ratio, P_o/P_u of the oriented membrane to the unoriented one. It is apparent that P for C_5H_{12} for

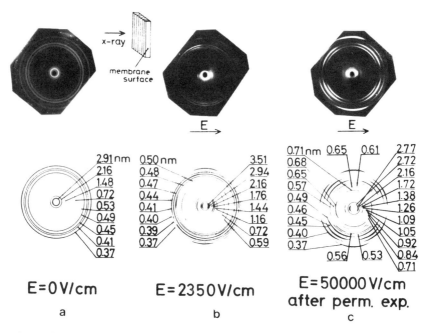

Figure 3.44 Wide-angle x-ray diffraction patterns of the PVC/CHOB (30/70) composite membrane prepared (a) without an imposed electric field and (b) with a 2350 V/cm field. The picture of (c) was taken after the permeation experiment under a field of 50,000 V/cm. The x-ray beam was directed along the sample surface (edge view).

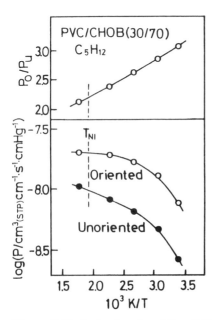

Figure 3.45 Arrhenius plot of P to C_5H_{12} for oriented and unoriented PVC/CHOB (30/70) composite membranes.

the oriented membrane is greater than that for the unoriented one. The magnitude of P increases and P_o/P_u decreases with increasing temperature. Similar tendencies were observed for iso-C_5H_{12} and neo-C_5H_{12}.

In the case of the unoriented composite membrane, the average channel dimension in the liquid-crystalline domain must be greater than that for the oriented one because of a looser packing of liquid-crystalline molecules, as illustrated in Fig. 3.41. This may lead one to expect that P for the unoriented membrane is greater than P for the oriented one. However, the experimental result is the reverse, as shown in Fig. 3.45. Consequently, the magnitude of P for the oriented and the unoriented composite membranes may be appreciably influenced by the different channel or path length required for diffusion through the composite membrane, as shown in Fig. 3.41. In other words, gas molecules diffuse along fairly straight paths through the oriented membrane but along tortuous paths through the unoriented one.

Since liquid-crystalline molecules have a tendency to form liquid-crystalline domains, molecular packing in these domains may be fairly tight even in the unoriented composite membrane. Therefore, the length of the diffusion path is more decisive for the permeability coefficient than for loose molecular packing. The temperature dependence of P for the unoriented composite membrane is slightly more marked than that for the oriented one, as shown in Fig. 3.45. This may be because thermal-molecular motion is more temperature dependent in the unoriented state with looser molecular packing.

According to the WAXD pattern of X-BPBB in the liquid-crystalline state, X-BPBB forms a cybotaitic nematic aggregation.[57] The relationship between the transition temperatures of X-BPBB and the intermolecular distance d is shown in Fig. 3.46. The phase transition temperatures from crystal to nematic (T_{KN}) and from nematic to isotropic (T_{NI}) decreased with increasing d due to reduction of the intermolecular interaction forces of X-BPBB.

Figure 3.47 gives an Arrhenius plot of the permeability coefficients of n-, iso-, and neo-C_6H_{14} and the permeability coefficient ratios among these isomers through the PVC/Br-BPBB (20/80) composite thin film. Above T_{KN} of Br-BPBB, P increased with decreasing the cross-sectional size of the permeant molecules. This trend suggests that permeation is predominantly governed by the diffusion process. The variation of P with temperature increases in the order n-, iso-, and neo-C_6H_{14} because permeation of the larger molecule is markedly reduced at lower temperatures, at which the free volume or the dimension of the channel becomes comparable to the dimensions of the permeant molecules. The permeability coefficient ratio of n-C_6H_{14} to neo-C_6H_{14} is the largest, and that of n-C_6H_{14} to iso-C_6H_{14} the smallest, in correspondence to the difference of the cross-sectional dimensions of these two isomers. A similar permeation trend was observed for the composite thin films containing other X-BPBB's.

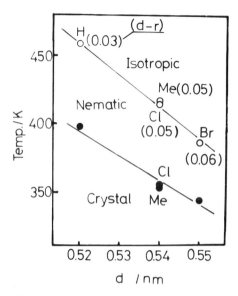

Figure 3.46 Relationship between the transition temperature and the intermolecular distance d of X-BPBB.

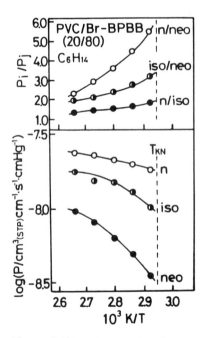

Figure 3.47 Arrhenius plot of P to n-, iso-, and neo-C_6H_{14} and the permeability coefficient ratios among these isomers for the PVC/Br-BPBB (20/80) composite membrane.

Figure 3.48 shows an Arrhenius plot of the diffusion coefficients D to n-, iso-, and neo-C_6H_{14} and the diffusion coefficient ratios among these isomers in the PVC/Br-BPBB (20/80) composite thin film. The temperature dependence of D for these isomers were similar to that of P. In the case of a dense membrane, P corresponds to the product of the solubility S and the diffusivity D; i.e., $P = SD$. Therefore, S was evaluated as P/D. The solubility coefficients for n-, iso-, and neo-C_6H_{14} calculated from Figs. 3.47 and 3.48 are comparable to each other. This result indicates that the main contribution to separation of hydrocarbon isomers through the PVC/X-BPBB membranes arises from the diffusion process rather than from the solution characteristics. When the appropriate size of an intermolecular channel for the diffusion of gas is formed by thermal fluctuation of the liquid-crystalline molecules in the composite thin film, gas molecules can diffuse in these channels. The larger penetrant isomers give a more marked temperature dependence of P and D than those for the smaller isomers, since P and D for larger isomers are more strongly temperature dependent than those for smaller isomers when the molecular diameter of the permeant isomer is comparable to the size of the intermolecular channel. Figures 3.47 and 3.48 indicate that the PVC/X-BPBB composite thin film can be a candidate for a novel molecular filtration membrane.

Figure 3.49 shows the intermolecular distance dependence of P to

Figure 3.48 Arrhenius plot of diffusivity to n-, iso-, neo-C_6H_{14} and the diffusivity ratios among these isomers for the PVC/Br-BPBB (20/80) composite membrane.

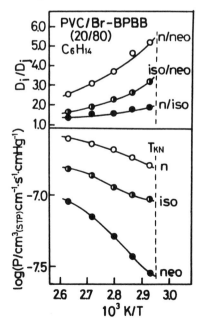

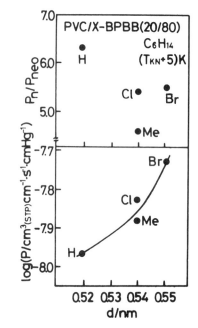

Figure 3.49 A plot of P for n-C_6H_{14} and the permeability coefficient ratios of n-C_6H_{14} to neo-C_6H_{14} for the PVC/X-BPBB (20/80) composite membranes against the intermolecular distance of X-BPBB at $(T_{KN} + 5)$ K.

n-C_6H_{14} and the permeability coefficient ratio of n- to neo-C_6H_{14} for the PVC/X-BPBB (20/80) composite membrane at $(T_{KN} + 5)$. Since an increase in the intermolecular distance induces an increase of the fraction of free volume or the channel dimension, both P and D of n-C_6H_{14} increase with increasing intermolecular distance. However, there is no apparent correlation between the permeability coefficient ratio of n-C_6H_{14} to neo-C_6H_{14} and the intermolecular distance of X-BPBB. Therefore, this is the subject for a future investigation of the relationships between permeation properties and the intermolecular distance by using liquid-crystalline materials that have an intermolecular distance difference of several Å.

3.4. Oxygen Enrichment through Polymer/LC/Fluorocarbon Monomer Ternary Composite Membranes

Figure 3.50 shows Arrhenius plots of permeability coefficients P for oxygen and nitrogen gas in the PVC/EBBA/(PFTA in L44 micelle) ternary composite membrane (PFTA-containing: curves 1 and 3) and the PVC/EBBA binary membrane (PFTA-free: curves 2 and 4). The broken lines represent the temperature regions of T_{KN} and T_{NI}. Over the whole temperature range studied here, values of P for both oxygen and

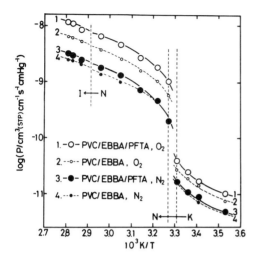

Figure 3.50 Arrhenius plots of permeability coefficients for oxygen and nitrogen gas in PFTA-containing (curves 1 and 3) and PFTA-free (curves 2 and 4) membranes.

nitrogen gas in the PFTA-containing composite membrane were greater than those in the PFTA-free one. In particular, an increase in oxygen permeability coefficient P_{O_2} was remarkable. An increase in P_{O_2} for the PFTA-containing composite membrane below T_{KN} indicates that the PFTA particles enhance the solubility of oxygen in the composite membrane. The degree of solubility of oxygen and nitrogen gas in FC monomer was evaluated by gas chromatography.[58] The solubility coefficient for oxygen gas in FC monomer is about 1.5 times that for nitrogen, as shown in Table 3.4, although FC monomer has considerable affinity for both gases. In addition, it has been reported that incorporation and release of gases in FC monomer are reversible, obeying Henry's law,[55,59] it is reasonable to consider that the additional increase in oxygen permselectivity in the PFTA-containing composite membrane may arise from the inherent solubility characteristics of FC monomer.

Figure 3.51 shows plots of P_{O_2}/P_{N_2} against P_{O_2} for PFTA-containing (curve 1), TPP-containing (curve 2), and FC-free (curve 3) composite membranes, together with gas-separation data for ordinary polymers for comparison.[60] An increase in P_{O_2} corresponds to an increase in measurement temperature. The value of P_{O_2} for the PVC/EBBA/FC ternary composite

Table 3.4 Solubilities of O_2 and N_2 in PFTA and L44 at 338 K

Compound	S_{O_2} [cm³ (STP)/100 ml]	S_{N_2} [cm³ (STP)/100 ml]	S_{O_2}/S_{N_2}
PFTA	35.4	22.8	1.56
L44	8.6	5.6	1.53

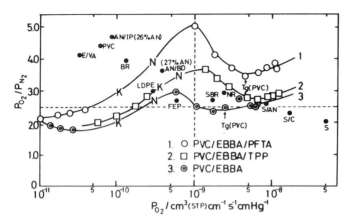

Figure 3.51 Variation of P_{O_2}/P_{N_2} with P_{O_2} for PFTA-containing (curve 1), TPP-containing (curve 2), and FC-free (curve 3) composite membranes, including reference data for various commonly used polymers.

membrane is higher than that of the FC-free membrane. A remarkable oxygen-enrichment effect was observed for the ternary composite membrane containing FC monomer. The order of magnitude of P_{O_2} was 10^{-9}–10^{-8} cm^3 (STP) cm^{-1} sec^{-1} (cm Hg^{-1}) and the ratio P_{O_2}/P_{N_2} was 3.5–4.0 (PFTA) and 2.8–3.5 (TPP) in the temperature range of the nematic or isotropic state of EBBA. This indicates that FC monomers play a role in enhancing the solubility of oxygen in the composite membrane surface. The ternary composite membrane exhibited unique behavior in that the ratio P_{O_2}/P_{N_2} increased with an increase of P_{O_2} above the glass transition temperature of the matrix polymer (marked by an arrow in Fig. 3.51). This trend may be caused by a desirable combination of the thermal-molecular motions from both matrix polymeric chains and liquid-crystalline molecules.[58,61,62] The unique relationship between P_{O_2}/P_{N_2} and P_{O_2} and their magnitudes make us expect that the composite thin film is practically applicable as an oxygen-enrichment membrane in the medical and engineering fields.

Although the mechanism of this unique behavior has not been clarified yet, we can speculate on the reason for the PFTA-free composite membrane as follows: The frequency at which the polymeric chains directly contact oxygen and nitrogen gas molecules on the upstream surface is significantly increased by the activated thermal-molecular motion of matrix PVC above T_g. In this case, the better oxygen selectivity of PVC than of the liquid-crystalline material may contribute to the improvement of the permeation characteristics above T_g.[63] Furthermore, in the case of the PFTA-containing composite membrane, we can interpret this unique trend from the inherent properties of the FC monomers. Although the magnitudes of solubility of both oxygen and nitrogen gas in FC monomers decrease with increasing temperature, the degree of decrease for oxygen is smaller than for nitrogen.[64] Consequently, this effect may contribute to an increase of P_{O_2}/P_{N_2} in spite of continuous increase in P_{O_2} above the T_g of PVC. There have been a few reports on successful results in attempts to improve the oxygen permselectivity by adding a material with good affinity for oxygen to the membrane.[65] The PVC/EBBA/PFTA ternary composite membrane can be an example of a characteristic permselective membrane for oxygen.

REFERENCES

1. J. Crank and G. S. Prak, Ed., *Diffusion in Polymers*, Academic Press, New York, N.Y. 1968.
2. P. Meares, *J. Am. Chem. Soc.* **76**, 3415 (1954).
3. A. S. Michaels, W. R. Vieth, and J. A. Barrie, *J. Appl. Phys.* **34**, 13 (1963).
4. T. Kajiyama, Y. Nagata, E. Maemura, and M. Takayanagi, *Chem. Lett.* **1979**, 679 (1979).
5. T. Kajiyama, *Membrane* **4**, 229 (1979); **6**, 265 (1981).
6. A. S. Michaels and R. B. Parker, Jr., *J. Polym. Sci.* **41**, 53 (1959).
7. R. M. Barrer and G. Skirrow, *J. Polym. Sci.* **3**, 549 (1948).
8. A. Aitken and R. M. Barrer, *Trans. Faraday Soc.* **51**, 116 (1955).
9. R. Y. M. Huang and P. J. F. Kanitz, *J. Appl. Polym. Sci.* **13**, 669 (1969).
10. C. A. Kumins, C. J. Rolle, and J. Roteman, *J. Phys. Chem.* **61**, 1290 (1957).
11. P. R. Seibel and F. P. McCandless, *Ind. Eng. Chem. Process. Dec. Dev.* **11**, 470 (1972).
12. E. C. Marthin, P. D. May, and W. A. McMahn, *J. Biomed. Mater. Res.* **5**, 53 (1971).
13. I. Langmuir, *J. Am. Chem. Soc.* **39**, 1848 (1917).
14. K. Blodgett, *J. Am. Chem. Soc.* **57**, 1007 (1935).
15. *Thin Solid Films* **132–134** (1986).
16. T. Oda, A. Takahara, M. Uchida, and T. Kajiyama, *Nippon Kagaku Kaishi* **1987**, 2163 (1987).
17. M. Sugi, *Thin Solid Films* **152**, 305 (1987).
18. K. Larsson, C. Nordling, K. Siegbahn, and E. Stenhagen, *Acta Chem. Scand.* **20**, 2880 (1966).
19. S. Hall, J. D. Andrade, S. M. Ma, and R. N. King, *J. Electron. Spectrosc. Relat. Phenom.* **17**, 181 (1979).
20. T. Kajiyama, N. Morotomi, S. Hiraoka, and A. Takahara, *Chem. Lett.* **1987**, 1737 (1987).
21. T. Kunitake, S. Tawaki, and N. Nakashima, *Bull. Chem. Soc. Jpn.* **53**, 3935 (1986).
22. T. Kunitake, Y. Okahata, and S. Yasunami, *J. Am. Chem. Soc.* **104**, 5547 (1982).
23. T. Kajiyama, A. Kumano, M. Takayanagi, Y. Okahata, and T. Kunitake, *Chem. Lett.* **1984**, 915 (1984).
24. A. Kumano, T. Kajiyama, M. Takayanagi, Y. Okahata, and T. Kunitake, *Ber. Bunsen-Ges. Phys. Chem.* **88**, 1216 (1984).

25. A. Kumano, T. Kajiyama, M. Takayanagi, Y. Okahata, and T. Kunitake, *Bull. Chem. Soc. Jpn.* **58**, 1205 (1985).
26. T. Kunitake, N. Higashi, and T. Kajiyama, *Chem. Lett.* **1984**, 717 (1984).
27. N. Higashi and T. Kunitake, *Chem. Lett.* **1986**, 105 (1986).
28. N. Higashi, T. Kunitake, and T. Kajiyama, *Macromolecules* **19**, 1362 (1986).
29. A. Takahara, N. Higashi, T. Kunitake, and T. Kajiyama, *Macromolecules* **21**, 2443 (1988).
30. V. Luzzati and F. Husson, *J. Cell. Biol.* **12**, 207 (1962).
31. A. Takahara, N. Morotomi, S. Hiraoka, N. Higashi, T. Kunitake, and T. Kajiyama, *Macromolecules* **22**, 617 (1989).
32. T. Kunitake, S. Tawaki, and N. Nakashima, *Bull. Chem. Soc. Jpn.* **56**, 3235 (1983).
33. T. Kunitake and N. Higashi, *J. Am. Chem. Soc.* **107**, 692 (1985).
34. T. Kajiyama, Y. Nagata, S. Washizu, and M. Takayanagi, *J. Membrane Sci.* **11**, 39 (1982).
35. T. Kajiyama, S. Washizu, and M. Takayanagi, *J. Appl. Polym. Sci.* **29**, 3955 (1984).
36. S. Hayashi, S. Ikuma, T. Kajiyama, and M. Takayanagi, *J. Polym. Sci., Polym. Phys. Ed.* **17**, 1995 (1979).
37. O. Ohara, C. Aso, and T. Kunitake, *J. Polym. Sci.* **11**, 1917 (1973).
38. T. Kunitake, T. Ochiai, and O. Ohara, *J. Polym. Sci. Polym. Chem. Ed.* **13**, 2581 (1975).
39. M. I. Davis and T. W. Muecke, *J. Phys. Chem.* **74**, 1104 (1970).
40. J. F. Chiang and C. F. Wilcox, Jr., *J. Am. Chem. Soc.* **73**, 2855 (1973).
41. F. A. Momany, L. M. Carruthers, R. F. McGuire, and H. A. Scheraga, *J. Phys. Chem.* **78**, 1595 (1974).
42. H. Sugimoto, M. Takayanagi, and T. Kunitake, *Polym. J.* **9**, 95 (1977).
43. G. L. Slonimskii, A. A. Askadskii, and A. I. Kitaigorodskii, *Polym. Sci. USSR* **12**, 556 (1970).
44. W. Frank, H. Goddar, and H. A. Stuart, *J. Polym. Sci. B.* **5**, 711 (1967).
45. H. Yasuda and V. Stannett, *Polymer Handbook*, second ed, J. Brandrup and E. H. Immergut, Eds., John Wiley & Sons, Inc., New York, N.Y., 1975, III-229.
46. S. Hayashi, T. Kajiyama, and M. Takayanagi, *Polym. J.* **13**, 443 (1981).
47. D. Chapman, R. M. Williams, and B. D. Ladbrooke, *Chem. Phys. Lipids* **1**, 445 (1967).
48. M. C. Block, L. L. M. van Deenen, and J. De Gier, *Biochim. Biophys. Acta* **433**, 1 (1976).
49. R. E. Kesting, *J. Macromol. Sci.-Chem.* **A4**, 665 (1970).
50. K. Maier and E. Scheuermann, *Kollid Z.* **171**, 122 (1960).
51. J. Crank, *The Mathematics of Diffusion*, Oxford Univ. Press, London, 1956.
52. S. Washizu, T. Kajiyama, and M. Takayanagi, *J. Chem. Soc. Jpn.* **1983**, 838 (1983).
53. T. Kajiyama, *J. Macromol. Sci.-Chem.* **A25**, 583 (1988).
54. M. Kawakami, H. Iwanaga, and S. Kagawa, *Bull. Chem. Soc. Jpn.* **54**, 869 (1981).

55. T. Kajiyama, S. Washizu, A. Kumano, I. Terada, and M. Takayanagi, *J. Appl. Polym. Sci., Appl. Polym. Symp.* **41**, 327 (1985).
56. T. Kajiyama, H. Kikuchi, and S. Shinkai, *J. Membrane Sci.* **36**, 243 (1988).
57. A. de Vries, *Mol. Cryst. Liq. Cryst.* **10**, 219 (1970).
58. T. Kajiyama, S. Washizu, and Y. Ohomori, *J. Membrane Sci.* **24**, 73 (1985).
59. T. Kajiyama, S. Washizu, and M. Takayanagi, *J. Appl. Polym. Sci.* **29**, 3955 (1984).
60. V. T. Stannett, W. J. Koros, D. R. Paul, H. K. Lonsdale, and R. W. Baker, *Adv. Polym. Sci.* **32**, 69 (1979).
61. T. Kajiyama, H. Kikuchi, I. Terada, and S. Shinaki, *Current Topics Polym. Sci.* **2**, 319 (1987).
62. Y. Ohmori and T. Kajiyama, *J. Chem. Soc. Jpn., Chem. Ind. Chem.* **1985**, 1897 (1985).
63. T. Nakagawa, H. B. Hopfenberg, and V. T. Stannett, *J. Appl. Polym. Sci.* **15**, 231 (1971).
64. M. K. Tham, R. D. Walker, Jr., and J. H. Modell, *J. Chem. Eng. Data* **18**, 385 (1973).
65. R. W. Baker, I. C. Roman, K. L. Smith, and H. K. Lonsdale, *Energy Technol.* **1982**, 505 (1982).

4 Design of Polymer Membranes for Gas Separation

Hisashi Odani and Toshio Masuda

1. Introduction
2. Fundamentals
   2.1. Definitions and Basic Equations
   2.2. Selectivity
   2.3. Concepts and Models
3. Membranes of Substituted Polyacetylenes for Gas Separation
   3.1. Polymer Synthesis
   3.2. Polymer Properties
   3.3. Gas Permeability and Permselectivity
4. Transport and Solution in a Disubstituted Polyacetylene: Permeation of Gases and Gas Mixtures in Poly[1-(trimethylsilyl)-1-propyne]
   4.1. Permeation Behavior of Pure Gases
   4.2. Sorption Behavior of Pure Gases and Vapors
   4.3. Structures of the Solvent-Cast Membrane
   4.4. Permeation Behavior of Gas Mixtures
   4.5. High-Resolution Solid-State ^{13}C NMR Spectroscopy

1. Introduction

The separation of gas mixtures employing polymer membranes has attracted much attention in recent years. A great amount of research, both scientific and technological, is now actively underway in many laboratories. This is due to the fact that the membrane separation technology is energy efficient and that process equipment is simple, compact, safe, and easy to control.

The major factors that are required for nonporous polymer membranes utilized in gas-separation applications are: (1) high flow rate(s) of the objective component(s) in the mixture, or vice versa (i.e., high permeability), (2) selectivity, or a high separation factor, and (3) mechanical and thermal strengths and chemical resistance to withstand environments imposed by the application. Unfortunately, it has been found generally that a reciprocal relation exists between permeability and selectivity. Accordingly, if one wishes to perform gas separation using a nonporous membrane of a glassy polymer, which bears satisfactory mechanical strength, a great reduction of the membrane thickness is needed to effect high gas fluxes. On the other hand, if we want to employ a rubbery polymer owing to its high permeability, efforts to prepare a composite membrane with tough and porous polymeric supports should be made to overcome its drawbacks, that is, to compensate for its unfavorable mechanical strength. From the point of view of the separation technology, the high productivity of the process is more appreciated than its high selectivity; that is, the high permeability of the polymer membrane is a preferable factor in gas-separation applications. Substituted polyacetylenes, which have been synthesized recently by Higashimura and one of the authors (T.M.), can be cast into durable membranes. Membranes of some substituted polyacetylenes exhibit high permeabilities for various gases despite their high glass transition temperatures, mostly higher than 200°C. Therefore, these substituted polyacetylenes are considered to be promising materials that can be used as single membranes for gas-separation applications.

In this chapter, we will first describe the fundamentals of the permeation behavior of gases and vapors through polymer membranes. The design of new polymeric materials for use in separation technology requires knowledge and an evaluation of many factors that affect the permeation behavior. Since we can find many reviews in the literature,[1] we confine ourselves here to a description of the basic equations and also concepts and models for permeation of gas mixtures.

In Section 3, the synthesis and various physical properties of the substituted polyacetylenes will be described with an emphasis on gas-permeation characteristics. An extremely high permeability of poly[1-(trimethylsilyl)-1-propyne] for various gases will be stressed there. The permeation behavior of gases and vapors and their mixtures in this disubstituted polyacetylene will be discussed in Section 4. Also, the behavior will be discussed in terms of microstructures and molecular motion of this polymer. Though the di-substituted polyacetylene exhibits unfavorable properties for use in gas-separation applications, information on microstructures and molecular motions of this polymer is considered to be very helpful in the development of new polymer membranes for gas separation with high productivity.

2. Fundamentals

2.1. Definitions and Basic Equations

The permeation of gases through a nonporous polymer membrane is governed by the coupled solution–diffusion mechanism.[2] The gas molecules dissolve in the surface layer on the ingoing side of the membrane, diffuse across the membrane in response to the concentration gradient, and evaporate from the other surface on the outgoing side.

Data usually obtained from permeation measurements are the amount of gas, Q_t, which has passed through a unit area of the membrane for a time t. A plot of Q_t versus t is called the *permeation curve*. As shown in Fig. 4.1, the permeation curve is convex toward the time axis at short times and then approaches a straight line as t increases. Permeation is in the steady state on the asymptotic linear portion of the permeation curve, since the

Figure 4.1 Typical permeation curve. The intercept on the time axis of the steady-state portin of the curve gives the time lag θ.

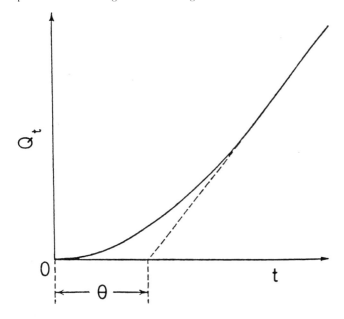

rate of permeation is independent of time there. In the steady state the concentration remains constant at all points of the membrane.

Theoretical relations necessary for the analysis of data obtained from isothermal permeation measurements with a membrane, where diffusion occurs effectively in one direction x, are obtained from solutions of the Fick diffusion equation,

$$\frac{\partial C}{\partial t} = \left(\frac{\partial}{\partial x}\right)\left[D\left(\frac{\partial C}{\partial x}\right)\right], \tag{1}$$

subject to appropriate initial and boundary conditions. Here, C is the concentration of penetrant and D is the *mutual diffusion coefficient* of the system.

At the beginning of a permeation experiment, the concentration is uniform everywhere in the membrane. This initial concentration is denoted by C_0. Then we have

$$C = C_0, \quad 0 < x < l, \, t = 0, \tag{2}$$

where l is the thickness of the membrane. The origin of x is taken at the surface on the ingoing side. The boundary condition for $x = 0$ is

$$C = C_\infty, \quad x = 0, \, t > 0, \tag{3}$$

where C_∞ denotes the equilibrium concentration corresponding to the pressure on the ingoing side, p_∞. The condition of constant surface concentration is fulfilled in the permeation of gases through polymer membranes in the rubbery state. If the swelling of the membrane during permeation is ignored, we may write

$$C = C_0, \quad x = l, \, t > 0. \tag{4}$$

Analytic solutions to Eq. (1) subject to these conditions can be obtained only for the special case in which D is independent of C.[3] When D is a constant, Q_t, under the condition of Eqs. (2)–(4) and $C_\infty \gg C_0 \approx 0$, is given by

$$\frac{Q_t}{lC_\infty} = \frac{D_0 t}{l^2} - \frac{1}{6} - \frac{2}{\pi^2}\sum_{n=1}^{\infty}\frac{(-1)^n}{n^2}\exp\left(-\frac{D_0 n^2 \pi^2 t}{l^2}\right), \tag{5}$$

where D_0 denotes constant D. As t goes to infinity, the steady state is approached and the exponential terms become negligibly small, so that the plot of Q_t versus t tends to a line

$$Q_t = \frac{D_0 C_\infty}{l}\left(t - \frac{l^2}{6D_0}\right). \tag{6}$$

The behavior was seen in Fig. 4.1.

The intercept on the time axis of the steady-state portion of the permeation curve is called the *time lag* θ (Fig. 4.1). From Eq. (6), θ is given by the relation

$$\theta = l^2 / 6D_0. \tag{7}$$

The steady-state permeation rate J_s is mathematically defined by

$$J_s = \lim_{t \to \infty} \frac{dQ_t}{dt}. \tag{8}$$

In view of Eq. (6), J_s is related to D_0 by the equation

$$J_s = D_0 C_\infty / l. \tag{9}$$

The relation between C_∞ and the corresponding pressure p_∞ is given by an expression of the form

$$C_\infty = S p_\infty, \tag{10}$$

where S is the solubility coefficient of the gas in the polymer. When Henry's law is obeyed, this is usually the case for systems of a simple gas and rubbery polymer, S is independent of concentration or pressure; that is $S = S_0$, a constant. Then J_s can be expressed as

$$J_s = D_0 S_0 p_\infty / l \tag{11}$$

or

$$P_0 \equiv D_0 S_0 = J_s l / p_\infty. \tag{12}$$

The product $P_0 \equiv D_0 S_0$ is referred to as the *permeability constant*. The value of P_0 is thus determined from the slope of the linear portion of the permeation curve.

For systems in which D is a function of concentration alone, that is, for Fickian system, we can deduce necessary relations to analyze permeation data without recourse to actual calculations. For a Fickian system, Eq. (9) is expressed as[4,5]

$$J_s(C_\infty) = \overline{D}(C_\infty) C_\infty / l. \tag{13}$$

Here $\overline{D}(C_\infty)$ is the quantity called the *integral diffusion coefficient* for the concentration C_∞ and is defined by

$$\overline{D}(C_\infty) = \frac{1}{C_\infty} \int_0^{C_\infty} D(C) \, dC. \tag{14}$$

For Fickian systems, nonlinear sorption isotherms are frequently observed. In these cases, the *solubility coefficient* is a function of concentration (or pressure); that is, $S = S(C_\infty)$. Using the concentration-dependent $S(C_\infty)$, we obtain from Eq. (13) the relation

$$\overline{P}(C_\infty) \equiv \overline{D}(C_\infty) S(C_\infty) = J_s(C_\infty) l / p_\infty. \tag{15}$$

$\overline{P}$ is referred to as the *(mean) permeability coefficient*.

The ordinary method of determining the solubility and diffusion coefficients is to measure the rates of absorption and desorption of penetrant in the polymer. If the diffusion coefficient is not constant or depends only on C, but depends on t or x, however, the value of D is not easily estimated from the absorption and desorption behavior. Even in this case, the substitution of the data of S and $\overline{P}$, which have been determined, respectively, from sorption and permeation measurements at various pressures, in Eq. (15) yields $\overline{D}$ as a function of concentration. This in turn allows the determination of the mutual diffusion coefficient D as a function C.[3,5] This procedure contains no approximation other than the neglect of the change in membrane thickness due to swelling and applies in circumstances in which D is time dependent or the condition of constant surface concentration is not obeyed; that is, absorption–desorption behavior is non-Fickian.

Alternatively, Eq. (15) provides a method of estimating the value of $\overline{P}$ not from permeability measurements but from sorption measurements. If the system concerned is in the rubbery state, or in the special case that the system is in the glassy state, data for both S and $\overline{D}$ are obtained from the sorption measurements, which were performed at various pressures. Using these data we may determine the value of $\overline{P}$ by employing the relation shown in Eq. (15). Good agreement between the values of $\overline{P}$ thus obtained and those determined directly from permeation measurements is reported.[6]

In the estimation of the time lag θ, the situation for concentration-dependent systems is more complicated. When D is a function of C only and the condition of constant surface concentration holds, θ is related to $D(C)$ by the equation[7]

$$\theta(C_\infty) = \frac{l^2 \int_0^{C_\infty} C\, D(C) \left(\int_C^{C_\infty} D(u)\, du \right) dC}{\left(\int_0^{C_\infty} D(C)\, dC \right)^3}. \tag{16}$$

This equation cannot be easily converted to permit the determination of $D(C)$ from θ. It is shown that an error involved in the determination of D from θ by employing the relation for constant D_0 [Eq. (7)] is small if $\overline{D}$ is used in place of D_0.[8]

2.2. Selectivity

When a gas mixture passes through nonporous polymer membranes, separation of the gas mixture occurs more or less during the permeation process. For example, in the permeation of O_2/N_2 mixtures through nonporous membranes, enrichment of O_2 in the permeated mixture was observed for a number of rubbery and glassy polymers,[9] although the enrichment mechanisms have not yet been well understood on the molecular level.

The selectivity of a polymer membrane for gas A relative to another gas B is characterized by the *ideal separation factor* α defined by

$$\alpha_{A/B} \equiv (Y_A / Y_B) / (X_A / X_B), \quad (17)$$

where Y_i is the mole fraction of the ith component in the outgoing-side receiver and X_i is the mole fraction of the ith component in the ingoing-side feed line. If the pressure on the outgoing side is much lower than that on the ingoing side, the separation factor in Eq. (17) can be approximated by[10]

$$\alpha_{A/B} = \overline{P}_A / \overline{P}_B. \quad (18)$$

By taking Eq. (15) into account, $\alpha_{A/B}$ is split into two parts:

$$\alpha_{A/B} = (\overline{D}_A/\overline{D}_B)(S_A/S_B). \quad (19)$$

The ratios $\overline{D}_A/\overline{D}_B$ and S_A/S_B are called the *mobility* (or *diffusivity*) *selectivity* and the *solubility selectivity*, respectively. The mobility selectivity is mostly governed by physical factors, such as molecular motions of the backbone chains and intersegmental packing in the polymer matrix. The solubility selectivity, on the other hand, is considered to be determined primarily on the basis of chemical interactions between gas molecules and the polymer. The separation behavior of gas mixtures employing nonporous membranes of various polymers has been discussed in terms of these selectivity terms.[11–15]

2.3. Concepts and Models

According to the state of penetrant–polymer systems, the permeation behavior of gas mixtures has been explained in terms of two different models, the dual-mode transport, or mobility, model and the free-volume model, although both models stem from the solution–diffusion concept.[1] If the system concerned is in the glassy state, the dual-mode transport model is conveniently used to analyze the permeation behavior of gas mixtures. On the other hand, the permeation behavior of a system of gas mixture and rubbery polymer is usually considered by the free-volume model. One can find many articles in the literature that show these models, respectively, explain the permeation behavior observed in the glassy and the rubbery regimes well, both being far from the glass transition region of a given system.[1,16–19]

2.3.1. The Dual-Mode Transport Model
This model has been formulated based upon the dual-mode sorption model.[16–19] For systems of a gas and a glassy polymer, sorption isotherms are frequently concave to the pressure axis, approaching linearity at higher pressures. The

sorption isotherm of this shape has been described by assuming that the penetrant dissolves by two component processes: ordinary dissolution in a continuous amorphous matrix (the concentration of this mode, C_D, is represented by Henry's law) and sorption in microvoids, or holes (the concentration C_H is represented by the Langmuir equation). The total concentration C of sorbed penetrant is then given by the expression of the dual-mode sorption model:

$$C = C_D + C_H = k_D p + C'_H bp/(1 + bp), \qquad (20)$$

where k_D is the Henry's-law constant, p is the pressure, and C'_H and b are, respectively, the Langmuir constant and the affinity constant.

By assuming that the penetrant species dissolved in the polymer matrix (the Henry's-law domain) and microvoids (the Langmuir domains) are both mobile and that their mobilities are characterized, respectively, by the diffusion coefficients D_D and D_H, the pressure dependence of the permeability coefficient P is expressed as

$$P = k_D D_D [1 + FK/(1 + bp)], \qquad (21)$$

where $F = D_H/D_D$ and $K = C'_H b/k_D$. In another derivation, the parameter F represents a fraction of the penetrant population dissolved in the Langmuir domains, which has the same mobility as the population dissolved in the Henry's-law domain. When D_D and D_H are constant, this formulation also gives the same expression for the pressure dependence of P as Eq. (21).[20] The dual-mode transport model predicts that P decreases with increasing penetrant pressure.

For cases involving only weak penetrant–penetrant and penetrant–polymer interactions, Koros et al. have shown that the dual-mode transport model can provide a description of the transport of a gas mixture in glassy polymers.[21] They have given the following expression for the permeability coefficient for component A in a binary-component feed A/B when the outgoing-side pressure can be approximated as zero:

$$P_A = k_D D_{DA} [1 + F_A K_A/(1 + b_A p_A + b_B p_B)]. \qquad (22)$$

Compared with Eq. (21), an additional term appearing in the denominator of the second term of Eq. (22) reflects the competition of the mixture for the limited microvoid capacity in the glassy polymer.[21,22] The permeability coefficient of either component is, therefore, predicted to be decreased due to the presence of the other component, which competes with it for sorption and pathways in the membrane. The predicted depression of the permeability coefficient for one component by the other has been substantiated in the permeation behavior of several systems of a binary gas mixture and a polymer in the glassy state.[19,23,24]

The dual-mode transport model was extended by Stern et al. to take into account the effects of polymer plasticization by penetrants.[25] Also, by assuming that both diffusion coefficients D_D and D_H are exponential functions of the concentration, Stern et al. have examined the plasticization effects on penetrant transport in the Henry's-law and the Langmuir domains separately.[26] An attempt to extend the formulation to analyze mixed gas permeation behavior has not been made yet.

2.3.2. The Free-Volume Model   The free-volume model, which has been developed by Stern et al. to represent permeation data of gas mixtures through membranes of rubbery polymers,[27,28] is based on an earlier model proposed by Fujita.[5,29] The validity of Fujita's model has been demonstrated for a number of systems of organic vapor and amorphous polymer in the rubbery state, which exhibit a rather strong dependence of the diffusion coefficient on penetrant concentration, and also on temperature.[1,4,8,16,18]

Stern et al. extended Fujita's free-volume theory of diffusion in polymers by taking into account of the effect of pressure on the free volume of the system. According to their treatment, the average fractional free volume of the system of penetrant and polymer f is related to a suitable standard state. This state is selected to be the fractional free volume of the pure polymer at a reference temperature T_s and a reference pressure p_s. Near the reference state, f at temperature T, pressure p, and volume fraction of penetrant v is represented by the relation

$$f(T, p, v) = f_s(T_s, p_s, 0) + \alpha(T - T_s) - \beta(p - p_s) + \gamma v, \tag{23}$$

where the thermal coefficient α, the pressure coefficient β, and the concentration coefficient γ are characteristic parameters. γ is a measure of the extent to which the polymer is plasticized by the penetrant. Equation (23) can be written in the form

$$f = f_0 + \gamma v, \tag{24}$$

where

$$f_0 = f_s + \alpha(T - T_s) - \beta(p - p_s). \tag{25}$$

Here f_o is the fractional free volume of the pure polymer at temperature T and pressure p. If the only pressure on the polymer membrane is that of the penetrant gas and the outgoing-side pressure is negligibly smaller than the ingoing-side one, then p in the above equations is equal to the latter pressure.

In the analysis of viscoelastic data of polymers by employing the WLF equation,[30] which explains well the temperature dependence of the shift factor for the time–temperature superposition in terms of the free-volume concept, T_s is usually taken as $T_g^o + 50°C$, where T_g^o is the glass transition temperature of a given polymer in the dry state. Also, Williams et al. have pointed out the fact that values of α for many polymers agree reasonably well with the difference between the thermal expansion coefficients above and below T_g^o.[30]

The extension of Eq. (24) to the case of transport of binary gas mixtures was made based on the following assumptions:

1. The solubilities of both components in the polymer are sufficiently low to obey Henry's law;
2. The effect of the two components on the free volume of the penetrant–polymer system is additive; that is, the second term on the right-hand side of Eq. (24) is represented by $\gamma_A v_A + \gamma_B v_B$, where v_A and v_B are, respectively, volume fractions of the components A and B, and γ_A and γ_B are the concentration coefficients; and
3. The molecules of the two penetrants do not differ much in size and shape, which implies, in turn, that the diffusion coefficients for them are of the same magnitude.

The permeability coefficient for component A is then given by the following relation when the outgoing-side pressure is negligibly small,

$$\ln P_A = C_A(T) + m_A(T)p_A + (B_{d_A}/B_{d_B})m_B(T)p_B, \qquad (26)$$

where $C_A(T)$, $m_A(T)$, $m_B(T)$, B_{d_A}, and B_{d_B} are parameters. The parameters $C_A(T)$ and $m_A(T)$, or $m_B(T)$, can be determined from permeability measurements with the pure component. The ratio B_{d_A}/B_{d_B} can be estimated approximately from the relation $B_{d_A}/B_{d_B} = (d_A/d_B)^2$, where d is the molecular diameter of the penetrant. Equation (26) predicts that P_A in the permeation of a binary gas mixture A/B is greater than that in permeation of pure gas A. Thus the extended free-volume model predicts an opposite pressure dependence of the permeability coefficient to that predicted by the dual-mode transport model. It has been reported that Eq. (26) describes satisfactorily the permeation behavior of binary gas mixtures through polyethylene[28] and the rubbery phase of styrene–butadiene block copolymer[31] membranes.

A new version of the free-volume theory has been developed by Vrentas and Duda.[32] The theory is based on the models of Cohen and Turnbull[33] and of Fujita[5,29] with the Bearman relation[34] between the mutual diffusion coefficient and the friction coefficient. The theory also employs the Flory–Huggins polymer solution theory[35] and the entanglement theory of Bueche.[36] Though this complex formulation explains the concentration dependence of the mutual diffusion coefficient for the solvent–polymer system well,[32] this model requires more parameters to be determined, which reduce the utility of the model. An extension of this model to the transport of gas–vapor mixtures has not been attempted yet.

3. Membranes of Substituted Polyacetylenes for Gas Separation

In 1983, 1-(trimethylsilyl)-1-propyne (MeC≡CSiMe$_3$; TMSP) was first polymerized with TaCl$_5$ and NbCl$_5$ into high polymers.[37] Poly(TMSP) proved extremely gas permeable; the oxygen permeability coefficient (P_{O_2}) of the polymer is ~ 7000 barrers [barrer = 1 × 10^{-10} cm^3 (STP) cm/cm^2 s cmHg], about ten times that of poly(dimethyl-siloxane), which had been the most gas-permeable polymer before poly(TMSP) emerged (see Section 3.3). Not only poly(TMSP) but also many other substituted polyacetylenes were synthesized in the past several years that also show fairly high gas permeability.

Herein the synthesis, properties, and gas-permeation behavior of substituted polyacetylenes are described as a case study of gas-permeable polymer membranes, although there have recently been further studies on new gas-permeable membranes (e.g., *p*-oligosiloxanyl polystyrene[38] and poly[4-(disilylmethyl)styrene][39]).

3.1. Polymer Synthesis

Substituted polyacetylenes have alternating double bonds along the main chain, and are usually synthesized by transition-metal catalyzed polymerization [Eq. (27)], where group-5 and -6 transition-metal (Nb, Ta, Mo, and W) catalysts are especially effective.[40–42] Typical examples of the synthesis are described here with emphasis on highly gas-permeable polymers such as poly(TMSP).

$$n\,R-C\equiv C-R' \xrightarrow{\text{catalyst}} \underset{R\quad R'}{+\!(C=C)\!+_n} \qquad (27)$$

3.1.1. Poly(TMSP) TMSP polymerizes with the pentachlorides and pentabromides of Nb and Ta[43] (Table 4.1). The polymerization is usually carried out in toluene at 80°C for 24 h. Both TaCl$_5$ and NbCl$_5$ quantitatively yield poly(TMSP)s, whose weight-average molecular weights ($\bar{M}_w$) are in the range 10^5–10^6. The former catalyst affords a polymer with higher molecular weight. The polydispersity ratios ($\bar{M}_w/\bar{M}_n$) of the polymers are usually 1.5–3.0.

Though the polymerization by TaCl$_5$ and NbCl$_5$ proceeds in various hydrocarbons and halogenated hydrocarbons, toluene is one of the most favorable solvents with respect to polymer yield and molecular weight. Poly(TMSP) can be quantitatively obtained in the temperature range 30–100°C after 24 h with both NbCl$_5$ and TaCl$_5$. The polymerization at 80°C is, however, the most practical when catalyst solubility, polymer molecular weight, and so on are taken into account.

Table 4.1 Synthesis of Poly(TMSP) and Analogous Polymers[a]

Monomer	Catalyst	Polymer Yield (%)	$\bar{M}_w/10^3$	Ref.
MeC≡CSiMe$_3$	TaCl$_5$	100	730	43
	NbCl$_5$	100	220	43
	TaCl$_5$–Ph$_3$Bi	100	4000	44
[b]	NbCl$_5$	100	140	45
MeC≡CSiMe$_2$CH$_2$SiMe$_3$	TaCl$_5$	100	1500	46
MeC≡CSiMe$_2$CH$_2$CH$_2$SiMe$_3$	TaCl$_5$–Ph$_4$Sn	60	400	46
MeC≡CSiMe$_2$-n-C$_6$H$_{13}$	TaCl$_5$–Ph$_3$Bi	70	1400	47
MeC≡CSiMe$_2$Ph	TaCl$_5$–Ph$_4$Sn	15	460	47
MeC≡CSiEt$_3$		25	1500	47
MeC≡CGeMe$_3$	TaCl$_5$	—	>100	48

[a]Polymerized in toluene at 80°C for 24 h; [M]$_0$ = 1.0 M, [Cat] = [Cocat] = 20 mM.
[b]Polymerized in cyclohexane; $\bar{M}_w/\bar{M}_n$ = 1.17.

A very high molecular weight poly(TMSP) can be obtained with a 1:1 mixture of TaCl$_5$ and Ph$_3$Bi.[44] The $\bar{M}_w$ reaches 4.0 × 10^6, which corresponds to the high intrinsic viscosity (13.2 dL/g in toluene at 30°C). This $\bar{M}_w$ is the highest among those of the substituted polyacetylenes ever known. The use of Ph$_3$Bi as a cocatalyst not only increases polymer molecular weight but also accelerates polymerization several times as compared with the polymerization by TaCl$_5$ alone. Among various organometallic cocatalysts, Ph$_3$Bi achieves the highest molecular weight. The NbCl$_5$–Ph$_3$Bi (1:1) catalyst produces a partly insoluble polymer.

The polymerization of TMSP by NbCl$_5$ in cyclohexane gives polymers with narrow molecular weight distributions (MWD) ($\bar{M}_w/\bar{M}_n$ ~1.2), irrespective of conversion.[45] The number-average molecular weight ($\bar{M}_n$) increases in direct proportion to conversion, and can be controlled in the range 1 × 10^4–20 × 10^4 by changing the monomer-to-catalyst ratio. This polymerization is a kind of living polymerization, although it is not known why the MWD becomes narrow only under this particular condition. Such poly(TMSP) with controlled $\bar{M}_n$ and MWD might be useful to design an excellent gas-separation membrane.

No catalysts other than TaCl$_5$- and NbCl$_5$-based ones induce polymerization of TMSP. For example, no reaction takes place in the presence of Mo and W catalysts under the same conditions as for the Ta and Nb catalysts. This implies that TMSP is sterically too crowded to polymerize with Mo and W catalysts, and that Ta and Nb catalysts are more useful for such monomers. Ziegler-type catalysts such as Fe(acac)$_3$–Et$_3$Al and Ti(OBu)$_4$–Et$_3$Al, which polymerize unsubstituted and monosubstituted acetylenes, do not effect TMSP polymerization either.

3.1.2. Analogues of Poly(TMSP) It is an interesting subject to synthesize analogues of poly(TMSP) and compare their gas permeability. MeC≡CSiMe$_2$CH$_2$SiMe$_3$ and MeC≡CSiMe$_2$CH$_2$CH$_2$Me$_3$, both of which have two silicon atoms, polymerize with TaCl$_5$ alone or its mixture with a suitable cocatalyst[46] (Table 4.1). The yields of polymers are good, and their $\bar{M}_w$ values are high enough (10^5–10^6).

MeC≡CSiMe$_2$–n-C$_6$H$_{13}$, MeC≡CSiMe$_2$Ph, and MeC≡CSiEt$_3$ afford fairly high molecular weight polymers by use of TaCl$_5$–cocatalyst systems.[47] The polymer yields are not always good, probably because of the long and bulky silyl substituents. Such steric effects are more prominent for bulkier monomers, MeC≡CSiMe$_2$–i-Pr, MeC≡CSiMe$_2$–t-Bu, and EtC≡CSiMe$_3$, from which no polymers have been obtained.

MeC≡CGeMe$_3$ polymerizes in the presence of TaCl$_5$ faster than TMSP to provide a film-forming, high molecular weight polymer quantitatively.[48] In contrast, neither MeC≡CCMe$_3$ nor MeC≡CSnMe$_3$ can be polymerized with Nb or Ta catalysts. Interestingly, poly(MeC≡CGeMe$_3$) does not dissolve in any organic solvents except CS$_2$.

3.1.3. Other Substituted Polyacetylenes Table 4.2 shows examples of the synthesis of high molecular weight, film-forming substituted polyacetylenes. It is worth noting that the monomers listed in Table 4.2 are methyl- and chlorine-containing disubstituted acetylenes or bulky-group-containing monosubstituted acetylenes. Thus the adequate seric crowding of monomers is essential to obtain high molecular weight, substituted polyacetylenes by using group-5 and -6 transition-metal catalysts.

MeC≡CPh, which is fairly sterically crowded, polymerizes with TaCl$_5$–n-Bu$_4$Sn.[40] When TaCl$_5$ alone is used as catalyst, the product polymer undergoes degradation into oligomers after all the monomer has been consumed. On the other hand, 2-alkynes, which are sterically less crowded, polymerize with Mo catalysts. 2-Alkynes give a large quantity of cyclotrimers in the presence of Nb and Ta catalysts.

Chlorine-containing disubstituted acetylenes such as ClC≡CPh and ClC≡C–n-C$_6$H$_{13}$ can be polymerized by Mo catalysts.[40] Especially, a catalyst obtained by uv irradiation of a CCl$_4$ solution of Mo(CO)$_6$ [Mo(CO)$_6$–hν catalyst] gives very high molecular weight polymers with $\bar{M}_w \sim 2 \times 10^6$. MeC≡CS–n-Bu, a sulfur-containing disubstituted acetylene, also polymerizes similarly with Mo catalysts.[49]

Among monosubstituted acetylenes, phenylacetylenes with bulky ortho-substituents (e.g., HC≡CC$_6$H$_4$–o-SiMe$_3$ and HC≡CC$_6$H$_4$–o-CF$_3$) form high molecular weight polymers in the presence of W catalysts (WCl$_6$–Ph$_4$Sn, W(CO)$_6$–hν, etc.).[50,51] With decreasing steric hindrance at the ortho position, the polymer molecular weight decreases accordingly (e.g., HC≡CC$_6$H$_4$–o-SiMe$_3$ > HC≡CC$_6$H$_4$–o-Me >> HC≡CPh); the $\bar{M}_w$ of poly(HC≡CPh) is no greater than 1×10^5. A few aliphatic, monosubsti-

Table 4.2 Synthesis of Various Substituted Polyacetylenes

Monomer	Catalyst	Method[a]	Polymer Yield (%)	$\bar{M}_w/10^3$	Ref.
MeC≡CPh	TaCl$_5$-n-Bu$_4$Sn	A	80	1500	40
MeC≡C–n-C$_5$H$_{11}$	MoCl$_5$-n-Bu$_4$Sn	B	60	~1000	40
ClC≡CPh	Mo(CO)$_6$-hν	C	80	2000	40
ClC≡C–n-C$_6$H$_{13}$		C	85	1800	40
MeC≡CS–n-Bu	MoCl$_5$-Ph$_3$SiH	A	60	180	49
HC≡CC$_6$H$_4$–o-Me	W(CO)$_6$-hν	C	100	800	50
HC≡CC$_6$H$_4$–o-SiMe$_3$	WCl$_6$-Ph$_4$Sn	B	90	1800	51
HC≡CC$_6$H$_4$–o-CF$_3$		D	100	1600	52
HC≡C–t-Bu	MoCl$_5$	B	100	~750	40
HC≡CCH(SiMe$_3$)–n-C$_5$H$_{11}$	MoCl$_5$-Et$_3$SiH	D	90	450	40

[a] Polymerization conditions for A are: in toluene, 80°C, 24 h, [M]$_0$ = 0.50 – 1.0 M, [Cat] = 10–30 mM. Conditions for B–D were modified as follows: B, 30°C; C, in CCl$_4$, 30°C; D, 0°C.

tuted acetylenes such as HC≡C–t-Bu and HC≡CCH(SiMe$_3$)–n–C$_5$H$_{11}$ also yield film-forming, high molecular weight polymers, whereas n-alkylacetylenes give only oligomers.[40]

3.2. Polymer Properties

The least requirement to obtain a free-standing polymer film is high enough molecular weight, and such a threshold molecular weight ($\bar{M}_w$) is roughly $>3 \times 10^5$ for substituted polyacetylenes. Apart from the film-forming property, mechanical properties and thermal stability are important factors for a polymer to serve as a gas-separation membrane.

3.2.1. Mechanical Properties[53]   Figure 4.2 shows stress–strain curves for typical substituted polyacetylenes. Polymers from aromatic disubstituted acetylenes (MeC≡CPh, ClC≡CPh, etc.) are hard and brittle, as indicated by the high Young's moduli (~2500 MPa), large tensile strengths (60–90 MPa), and small elongations at break (<5%). In contrast, the aliphatic counterparts like poly(TMSP) and poly(MeC≡C–n-C$_5$H$_{11}$) are soft and ductile; the Young's moduli are smaller (600–800 MPa), the tensile strengths are ~40 MPa, and the elongations at break reach ~70%. Further, poly(MeC≡C–n-C$_7$H$_{15}$) and poly(ClC≡C–n-C$_8$H$_{17}$), which have long n-alkyl groups, are very soft and ductile (Young's moduli 50–250 MPa, tensile strengths ~10 MPa, elongations at break 200–400%).

According to dynamic–viscoelastic measurements, the glass transition temperatures (T_g) of poly(MeC≡CPh) and poly(ClC≡CPh) are in the vicin-

Figure 4.2 Stress–strain curves for substituted polyacetylenes at 25°C. Strain rate: 86%/min.

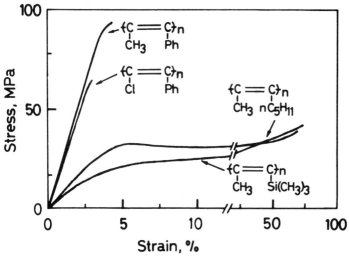

ity of 200°C. The T_g of poly(TMSP) is estimated to be above 200°C. The value of poly(MeC≡C–n-C$_5$H$_{11}$) appears somewhat lower (~180°C). Thus, most polyacetylenes have T_g's around 200°C that do not strongly depend on the kind of substituents. This contrasts with the T_g's of vinyl polymers, which are usually much lower and vary widely with the kind of substituents. It is thought that the T_g of polyacetylenes is mainly governed by the alternating double-bond structure of the main chain.

3.2.2. Thermal Stability[54] The temperature at which a polymer begins to lose weight in thermogravimetric analysis (TGA) is a measure of the thermal stability of the polymer. Though the TGA of polymers is carried out either in air or in an inert atmosphere, the data observed in air should be more important because polymers are usually used under this condition.

Figure 4.3 depicts TGA curves of several substituted polyacetylenes measured in air. Poly(MeC≡CPh) and poly(ClC≡CPh) maintain their weight up to ~300°C. The starting temperature of weight loss of poly(TMSP) is ~280°C. In contrast, poly(MeC≡C–n-C$_5$H$_{11}$) begins to lose weight as low as ~200°C. Namely, the thermal stability of film-forming, disubstituted acetylene polymers decreases in the order: aromatic > heteroatom-containing > aliphatic.

The thermal stability of a polymer in air can also be estimated by its molecular weight change on heat treatment. In the case of substituted poly-

Figure 4.3 TGA curves for substituted polyacetylenes in air. Heating rate: 10°C/min.

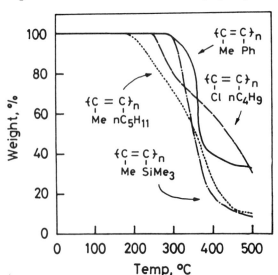

acetylenes, it is convenient to perform the heat treatment in air at 120°C for 20 h. Under this condition, poly(MeC≡C-n-C$_5$H$_{11}$) remarkably degrades into oligomers ($\bar{M}_w$ = 10^5 to 10^3). The $\bar{M}_w$ of poly(TMSP) decreases to some extent. In contrast, poly(MeC≡CPh) and poly(ClC≡CPh) do not suffer an $\bar{M}_w$ decrease at 120°C. These results correspond to the TGA data shown above. None of these polymers undergoes degradation in vacuo at 120°C. The products of the polymer degradation in air contain appreciable amounts of oxygen. These findings manifest that the oxidation process is essential to the thermal degradation.

It has been found that some antioxidants are useful to protect poly(TMSP) from air oxidation.[55,56] The time to failure of poly(TMSP) can be determined by a TGA weight loss of 2%, and its 2-year service life temperature is estimated to be 52°C. The addition of antioxidants at a 2% level raises the 2-year service life temperature [e.g., up to 103°C with Irganox 1010 (Ciba-Geigy)].

3.2.3. Other Properties[40]

General properties of poly(TMSP) are described: It is a colorless solid. This agrees with the fact that its uv–visible spectrum exhibits only a small absorption in the uv region (λ_{max} = 235 nm, ε_{max} = 3400 M^{-1} cm^{-1}) and none above 300 nm. This polymer is amorphous according to x-ray diffraction analysis. The polymer synthesized with TaCl$_5$ dissolves in toluene, xylene, benzene, cyclohexane, heptane, CCl$_4$, CHCl$_3$, and THF, but does not dissolve in (CH$_2$Cl)$_2$, 1,4-dioxane, anisole, diethyl ether, ethyl acetate, acetone, acetic acid, nitrobenzene, DMF, or DMSO. The polymer obtained with NbCl$_5$ is somewhat less soluble (insoluble in heptane). Casting poly(TMSP) from toluene solution provides a colorless, transparent, uniform, free-standing film. Poly(TMSP) is an electrical insulator with a specific conductivity of 1 × 10^{-17} S cm^{-1}. No unpaired electrons are found in this polymer by ESR. The insulating and nonparamagnetic properties are due to the twisted conformation of the main chain carrying bulky Me$_3$Si groups.

Most of the above-stated properties of poly(TMSP) are common among many substituted polyacetylenes listed in Tables 4.1 and 4.2. Polymers from disubstituted acetylenes and aliphatic monosubstituted acetylenes are colorless or pale yellow. Polymers of ortho-substituted phenylacetylenes are, in contrast, colored dark brown or purple. They are all amorphous according to x-ray diffraction. These polymers are usually soluble in low-polarity solvents such as toluene and CHCl$_3$, and hence films can be easily fabricated by solution casting. They are thermally fairly stable in air, and electrically insulating. These properties present a striking contrast to those of polyacetylene [e.g., lustrous (film), crystalline, insoluble, and semiconducting]. Thus, substituted polyacetylenes constitute a novel category of polymers that clearly differ from polyacetylene.

3.3. Gas Permeability and Permselectivity

3.3.1. Effect of Substituents

In Table 4.3 are listed the oxygen permeability coefficients, P_{O_2}, and the ratios of permeability coefficients for oxygen and nitrogen, P_{O_2}/P_{N_2} (ideal separation factor), so far determined for substituted polyacetylenes at 25°C. (Unless the pressure-independent nature of the gas-permeability coefficient is confirmed, the symbol $\bar{P}$, which stands for the mean permeability coefficient defined in the preceding section, should be used. For the simplicity of presentation, however, we use hereafter the symbol without the overbar.) The P_{O_2} values are in the range ~6000–5 barrers. For the sake of comparison, the values for conventional oxygen-permeable polymer membranes are cited (25°C):[17,61,62] poly(dimethylsiloxane) P_{O_2} = 600 barrers (P_{O_2}/P_{N_2} = 2.0); poly(4-methyl-1-pentene) 32 (—); natural rubber 23 (2.3); poly(oxy-2,6-dimethylphenylene) 15 (5). Thus, substituted polyacetylenes have high P_{O_2} values among various polymers. Their high gas permeability is attributable to their stiff main chain and bulky substituents.

The P_{O_2} of poly(TMSP) is roughly 3000–6000 barrers, extremely high (actually, this value fluctuates in a wider range of 3000–10,000 barrers, depending on research groups and measuring conditions.[37,57,58,63,64]) Further, several substituted polyacetylenes show high P_{O_2} in the order of 10^2–10^3 barrers. The P_{O_2} of many substituted polyacetylenes are in the range 10–100 barrers, while a few polyacetylenes have P_{O_2} values of 1–10 barrers. As seen in Table 4.3, the polymers that have P_{O_2} above 100 barrers all bear bulky, rigid pendants such as Me_3Si, Me_3Ge, and t-Bu groups. The polyacetylenes whose P_{O_2} are 10–100 barrers carry either a long n-alkyl group or two kinds of substituents among silyl, n-alkyl, and phenyl groups. The polymers with P_{O_2} lower than 10 barrers are featured by the presence of phenyl group as principal pendant.

Table 4.4 shows permeability coefficients (P) of substituted polyacetylenes for six gases. The polymers in Table 4.4 have been classified into three groups according to both the kind of substituents and their gas-permeation behavior. Those of the first, second, and third groups carry a bulky substituent, a long n-alkyl group, and a phenyl group, respectively. Interestingly, the polyacetylenes of the first group exhibit large P values for all gases. Although among the four polymers of the first category, three in common have a silyl group, the fact that poly(HC≡C–t-Bu) is also very permeable indicates that a silyl pendant group is not a requisite to high gas permeability. The P values for the second group are medium in magnitude. This means that long n-alkyl groups do not greatly affect the gas permeability of substituted polyacetylenes. The polymers of the third group show rather small P values. This suggests that phenyl groups stack on one another, which disfavors gas permeation.

As seen in Fig. 4.4, the apparent activation energies E_P for gas permea-

Table 4.3 Oxygen Permeability Coefficients (P_{O_2}) and Ratios P_{O_2}/P_{N_2} of Substituted Polyacetylenes[a]

$$-(C=C)_n-$$
$$||$$
$$RR'$$

R	R'	P_{O_2}[b] (barrer)	$\dfrac{P_{O_2}}{P_{N_2}}$	Ref.
Me	SiMe$_3$	3000–6000	1.7	37,57,58
Me	GeMe$_3$	1800[c]	1.5	48
Ph	C$_6$H$_4$–m-SiMe$_3$	1200	2.0	59a
Ph	C$_6$H$_4$–p-SiMe$_3$	1100	2.1	59a
Me	SiEt$_3$	860	2.0	47
Me	SiMe$_2$Et	500	2.2	58
H	t-Bu	130	3.0	57
Me	SiMe$_2$–n-Pr	100	2.8	58
Me	S–n-Bu	79	4.4	49
H	C$_6$H$_4$–o-SiMe$_3$	78	3.3	57
Me	SiMe$_2$CH$_2$SiMe$_3$	75	3.6	57
Me	SiMe$_2$CH$_2$CH$_2$SiMe$_3$	50	3.6	57
Me	S–n-C$_8$H$_{17}$	50	2.6	49
Cl	n-C$_8$H$_{17}$	47	2.9	57
Me	S–n-C$_{10}$H$_{21}$	46	2.9	49
Me	C$_6$H$_4$–m-SiMe$_3$	40	3.1	59b
Me	S–n-C$_6$H$_{13}$	38	2.7	49
Me	n-C$_7$H$_{15}$	35	2.5	57
Cl	n-Bu	35	3.5	57
Me	n-C$_5$H$_{11}$	34	2.5	60
Cl	n-C$_6$H$_{13}$	32	2.9	57
Me	S–Et	30	3.3	49
H	CH(n-C$_5$H$_{11}$)SiMe$_3$	27	3.1	57
H	CH(n-C$_7$H$_{15}$)SiMe$_3$	27	3.0	40
H	C$_6$H$_4$–o-CF$_3$	25	3.4	57
H	CH(n-C$_3$H$_7$)SiMe$_2$–n-C$_6$H$_{13}$	19	3.0	57
Me	SiMe$_2$–n-C$_6$H$_{13}$	18	4.2	57
Ph	n-C$_6$H$_{13}$	14	2.5	57
Et	Ph	12	2.7	57
H	CH(n-C$_3$H$_7$)SiMe$_2$Ph	9.5	3.8	57
H	C$_6$H$_4$–o-Me	8.1	2.7	57
Me	Ph	6.3	2.9	57
Cl	Ph	5.1	5.1	57

[a] Measured at 25°C.
[b] 1 barrer = 1×10^{-10} cm^3(STP) cm/(cm^2 s cm Hg).
[c] Estimated as 0.4 times an average value, 4500, of poly(TMSP); cf. Ref. 48.

tion through poly(TMSP) are more or less negative; that is, the lower the temperature, the more permeable the membrane. In sharp contrast, the E_P values of all the conventional polymers are positive except those for readily condensable gases. The E_P values of poly(MeC≡CPh) and poly(MeC≡C–n-C$_7$H$_{15}$) are similar to those of poly(dimethylsiloxane). Less gas-permeable polymers usually have much larger E_P values.

Table 4.4 Gas Permeability Coefficients (P) of Substituted Polyacetylenes[a]

$-(CR=CR')_n-$		\multicolumn{6}{c}{P (barrer)[b]}				
R	R'	He	H_2	O_2	N_2	CO_2	CH_4
\multicolumn{8}{l}{R' = a bulky substituent}							
Me	$SiMe_3$	2200	5200	3000	1800	19000	4300
Me	$SiMe_2CH_2SiMe_3$	180	270	75	21	310	45
H	t-Bu	180	300	130	43	560	85
H	C_6H_4–o–$SiMe_3$	170	290	78	24	290	38
\multicolumn{8}{l}{R' = a long n-alkyl group}							
Me	n-C_7H_{15}	48	76	35	14	130	40
Cl	n-C_8H_{17}	43	76	47	16	170	46
Cl	n-C_6H_{13}	71	66	32	11	130	33
Me	S–n-C_8H_{17}	57	91	50	19	180	48
\multicolumn{8}{l}{R' = a phenyl group}							
Me	Ph	30	43	6.3	2.2	25	2.8
Et	Ph	40	57	12	4.5	40	4.4
Cl	Ph	23	29	5.1	1.0	23	1.3
H	C_6H_4–o–Me	29	39	8.1	3.0	15	3.0

[a]Measured at 25°C, Refs. 57 and 49.
[b]1 barrer = 1×10^{-10} cm^3(STP) cm/(cm^2 s cm Hg).

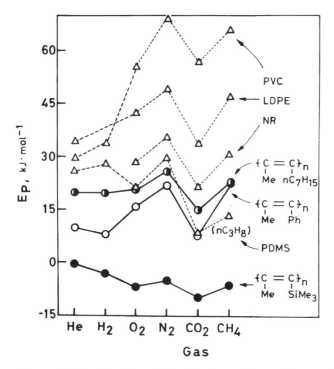

Figure 4.4 E_P values for substituted polyacetylenes and conventional polymers. (Reproduced from Ref. 57. Copyright 1988 Butterworth Co. Ltd.)

3.3.2. Solubility Coefficient and Diffusion Coefficient As described in the preceding section, the approximate value of the (integral) diffusion coefficient can be evaluated from the time lag using Eq. (7), which in turn allows us to evaluate the solubility coefficient S by employing the relation shown in Eq. (15), that is, by dividing the (mean) permeability coefficient by the diffusion coefficient.

Figure 4.5 plots S versus D for oxygen permeation through substituted polyacetylenes.[65] (Hereafter, the overbar to denote the integral diffusion coefficient is removed for simplicity of presentation.) Because $P = SD$, the P value increases as a point goes up and to the right in Fig. 4.5. The polyacetylenes (points 1–3) with a bulky substituent have large S and D values, which results in large P values (cf. Table 4.3). Points 4–6 for the polyacetylenes with a long n-alkyl group lie below those with a bulky substituent, showing that their smaller P values are due to the smaller S. In contrast, the polyacetylenes with a phenyl group (points 7–9) are plotted on the left of the polyacetylenes with a bulky substituent. This means that their small P values stem from their small D values.

Not only for conventional polymers but also for substituted polyacetylenes, do good linear relationships hold between log S and the boiling point of the permeant gas (Fig. 4.6). As the boiling point of the gas increases, the

Figure 4.5 Solution versus diffusion plots for the O_2 permeation through substituted polyacetylenes at 25°C. (Reproduced from Ref. 65. Copyright 1988 The Membrane Society of Japan.)

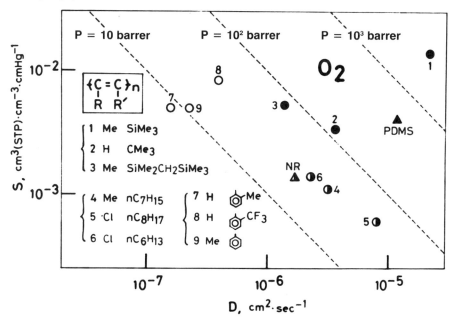

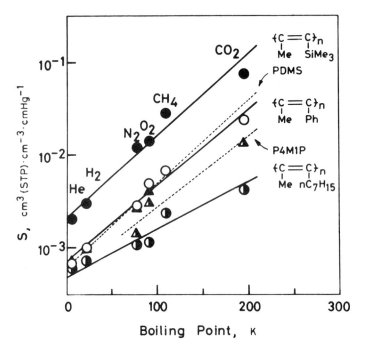

Figure 4.6 Relationship between boiling point of gases and solubility coefficient at 25°C. (Reproduced from Ref. 57. Copyright 1988 Butterworth Co. Ltd.)

S value increases for all these polymers. Interestingly, the S value of poly(TMSP) is the largest with every gas. The S values of poly(MeC≡CPh) follow, while those of poly(MeC≡C-n-C$_7$H$_{15}$) are small. These results suggest that the affinity for gases decreases in the order of polymers having a bulky substituent, phenyl group, and a long n-alkyl group.

Figure 4.7 depicts the relationship between D and the molecular diameter of the permeant gas. The log D of polymers decreases linearly with increasing molecular diameter of the gas. Regarding the substituent effect of polyacetylenes, the D value decreases with every gas in the order of poly(TMSP) > poly(MeC≡C-n-C$_7$H$_{15}$) > poly(MeC≡CPh). The very large D values of poly(TMSP) are attributable partly to the presence of molecular-scale voids in the polymer. In contrast, the relatively large D values of poly(MeC≡C-n-C$_7$H$_{15}$) are probably due to the flexibility and mobility of the n-heptyl group. It is obvious that the rigid phenyl group is unfavorable for the diffusion of large gas molecules.

It was described previously that the E_P values of poly(TMSP) are very small. According to the solution–diffusion mechanism, $E_P = \Delta H_S + E_D$, where ΔH_S is the heat of solution and E_D is the apparent activation energy for diffusion. In the oxygen permeation through poly(TMSP), $E_P = -6.7$ kJ mol^{-1}, $\Delta H_S = -11.7$ kJ mol^{-1}, and $E_D = 5.0$ kJ mol^{-1}. The ΔH_S values

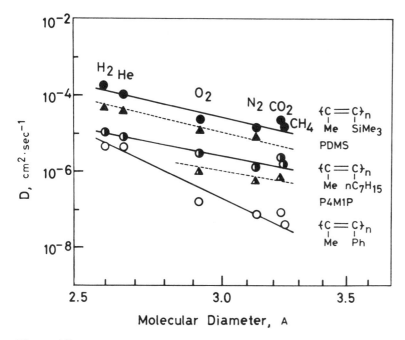

Figure 4.7 Relationship between molecular diameter of gases and diffusion coefficient at 25°C. (Reproduced from Ref. 57. Copyright 1988 Butterworth Co. Ltd.)

for various polymers stay in a relatively narrow range of -20 to 0 kJ mol^{-1}, whereas the E_D values span a wide range from 5 to 60 kJ mol^{-1}. Consequently the small E_P of poly(TMSP) originates from its small E_D, which seems associated with the high T_g and the microvoid-containing loose structure of this unique polymer. The structure of the poly(TMSP) membrane will be discussed in the following.

3.3.3. Poly(TMSP) and Poly(dimethylsiloxane) The P, S, and D values of poly(TMSP) at 25°C are, reportedly, 3000 barrers, 14 × 10^{-3} cm^3(STP) cm^{-3} cm Hg^{-1}, and 220 × 10^{-7} cm^2 s^{-1}, respectively,[57] whereas those of poly(dimethylsiloxane) at 20°C are 350 barrers, 1.9 × 10^{-3} cm^3(STP) cm^{-3} cmHg^{-1}, and 190 × 10^{-7} cm^2 s^{-1}.[66] Namely, the S of poly(TMSP) is several times larger than that of poly(dimethylsiloxane), while the D of the former is slightly larger than that of the latter. Poly(TMSP) is, therefore, an extreme example of glassy polymers that show large P values. Not only the P but also S and D of poly(TMSP) are the largest among the values of all the existing polymers.

Many differences in properties can be pointed out between these two highly gas-permeable polymers. For instance, the T_g of poly(TMSP) is higher than 200°C, and it is in the glassy state at room temperature. In contrast, the T_g of poly(dimethylsiloxane) is -127°C, and hence it is in the rubbery state at room temperature. A thin tough film can be easily fabricated from the former by solution casting, which is difficult for the latter. Gas permeation through poly(dimethylsiloxane) is interpretable simply in terms of the solution–diffusion mechanism. In contrast, the dual-mode mechanism involving Langmuir adsorption[63,67] or more complicated mechanisms[67] hold in some cases with poly(TMSP).

3.3.4. Modifications of Poly(TMSP) As will be described in detail later, poly(TMSP) has two disadvantages when it is applied to a membrane for oxygen enrichment, that is, the small separation factor below 2 and the decrease in P_{O_2} with time. For the purpose of improving these disadvantages, the following modifications of poly(TMSP) have been examined.

It has been reported that fluorine treatment of poly(TMSP) membrane is useful to enhance its permselectivity.[64] The fluorine treatment is carried out by exposing a poly(TMSP) membrane, for instance, in an atmosphere of fluorine and nitrogen ($F_2 = 1\%$) for 15 min at room temperature. According to the ESCA analysis of the membrane surface thus treated, fluorine and oxygen are introduced, while the Si content decreases appreciably. The fluorine treatment decreases the P_{O_2} value of poly(TMSP) to one-tenth the original value (Table 4.5). Quite interestingly, however, the P_{O_2}/P_{N_2} ratio increases up to 4.6. This value means that the oxygen in air is enriched to about 50%. In general, the permeability to gases, especially large molecules like CH_4, CO_2, and N_2, decreases by fluorine treatment, resulting in an enhancement of permselectivity. This is attributable to the formation of a dense structure on the membrane surface.

Plasma polymerization has been employed to prepare a poly(TMSP) membrane that has a denser structure and may show a higher permselectivity.[68] The polymer obtained hardly possesses a $C=C$ bond, but comprises a cross-linked structure. The P_{O_2} value (20 barrers) of this polymer membrane is much smaller than the value of the polymer produced catalytically, while its P_{O_2}/P_{N_2} (2.5) is somewhat larger.

Table 4.5 Gas-Permeation Behavior of Fluorinated Poly(TMSP)[a]

Poly(TMSP)		Untreated	Fluorinated
Membrane thickness (μm)		141	97
P (barrer)	Ne	6800	5100
	H_2	16000	6200
	O_2	10000	970
	N_2	6800	210
	CO_2	32000	3700
	CH_4	16000	120
P_x/P_y	O_2/N_2	1.5	4.6
	He/CH_4	0.43	43
	CO_2/CH_4	2.0	31

[a]Fluorinated under a F_2/N_2 (F_2 1%) atmosphere for 15 min, Ref. 64.

In an attempt to suppress the decrease of the P_{O_2} of poly(TMSP) membrane with time, the introduction of poly(dimethylsiloxane) grafts onto poly(TMSP) has been examined.[69] Such graft copolymers are prepared, for example, according to

$$\begin{array}{c} \mathrm{+\!(C=C)\!\!\!\!\!\!\!\!-}_{\overline{m}} \\ \phantom{\mathrm{+\!(}}|| \\ \phantom{\mathrm{+\!(}}\mathrm{Me}\mathrm{SiMe_3} \end{array} \xrightarrow[\text{(2) ClSiMe}_2\text{CH}_2\text{CH}_2(\text{SiMe}_2\text{O})_m\text{SiMe}_3]{\text{(1) }n\text{-BuLi}}$$

$$\begin{array}{c} \mathrm{+\!(C=C)\!\!\!\!\!\!\!\!-}_{\overline{m\text{-}x}}\mathrm{(C=C)\!\!\!\!\!\!\!\!-}_{\overline{x}} \\ \phantom{\mathrm{+\!(}}||\phantom{\mathrm{(}}|| \\ \phantom{\mathrm{+\!(}}\mathrm{Me}\mathrm{SiMe_3}\phantom{\mathrm{(}}\mathrm{SiMe_3} \\ \phantom{\mathrm{+\!(C=C)\!\!\!\!\!\!\!\!-}}\mathrm{CH_2SiMe_2CH_2CH_2(SiMe_2O)}_m\mathrm{SiMe_3} \end{array} \qquad (28)$$

It is hoped that this polymer will possess both an availability of a thin membrane like poly(TMSP) and invariable permeability like poly(dimethylsiloxane). As seen in Fig. 4.8, the graft copolymer membrane whose

Figure 4.8 Effect of graft content on the P_{O_2} and P_{O_2}/P_{N_2} of poly(TMSP)–graft-poly-(dimethylsiloxane). (Reproduced from Ref. 69. Copyright Koubunshi Kankoukai.)

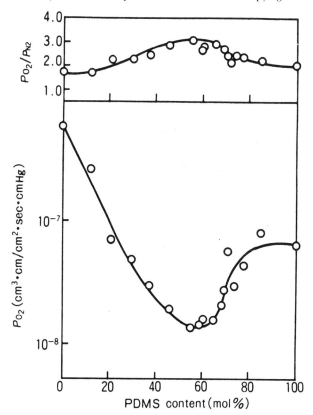

poly(dimethylsiloxane) content is 70% exhibits P_{O_2} and P_{O_2}/P_{N_2} values similar to those of poly(dimethylsiloxane). This membrane shows no decrease in P_{O_2} with time, and is tough enough even if thin.

4. Transport and Solution in a Disubstituted Polyacetylene: Permeation of Gases and Gas Mixtures in Poly[1-(trimethylsilyl)-1-propyne]

In the preceding section, we have stressed the extremely high permeabilities for various gases through glassy poly(TMSP) membrane. However, the high permeability coefficient for O_2, for instance, decreases by about an order of magnitude after treating the membrane at 100°C for several hours.[58] The result suggests that molecular-scale gaps or microvoids exist in the solvent–cast-polymer matrix and that they collapse during the thermal treatment. In the first part of this section, the permeation and solution behavior of gases and vapors in poly(TMSP) membrane will be described. The transport and solution behavior of penetrant in polymer solid is sensitive to the polymer structure on a scale comparable to the size of penetrant molecules, and studies of the behavior of many polymers have contributed much to our understanding of the microstructure and the molecular motions of the polymers as well as that of transport mechanisms.[4,70]

In the second part, the permeation behavior of gas mixtures will be shown and discussed in terms of the two leading models of gas transport, that is, the dual-mode transport and the free-volume models described in Section 1. Also, the permeation behavior will be interpreted using information on molecular motions of poly(TMSP), which has been revealed by ^{13}C and ^{29}Si NMR studies.

4.1. Permeation Behavior of Pure Gases

In order to confirm the nonporous nature of the solvent-cast poly(TMSP) membrane, the permeation behavior was studied for O_2, N_2, and five inert gases (He, Ne, Ar, Kr, Xe). Permeation data for these gases showed that no correlation exists between P and the reciprocal of the square root of the molecular weight of the gas. This indicated that the permeation behavior does not obey Graham's law of effusion. Thus it is concluded that the poly(TMSP) membrane is nonporous.

The values of P for these gases were almost constant in the pressure region below 1 atm. For a larger penetrant molecule, such as isobutane and acetone, P increased with increasing p in the whole pressure region studied, $0 < p \leq 1$ atm. As has been described in Section 1, the pressure dependence of P observed for these penetrants does not conform to that predicted by the dual-mode transport model [Eq. (21)], but agrees with that predicted by the free-volume model [Eq. (26) with $p_B = 0$], although the systems concerned are in the glassy state at the temperature of the experiments. The sorption isotherm for isobutane, however, exhibits a shape that can be analyzed well in terms of the dual-mode sorption model[67] (Fig. 4.10).

A noticeable physical aging effect on P was observed as shown in Fig. 4.9[67,71] This figure shows plots of P for isobutane against total time under vacuum at 30, 50, and 70°C. The applied pressure is 30 cm Hg, and the origin of time is taken at a time when the sample membrane has been brought out from leaching-liquid methanol. The leaching was made to remove traces of casting solvent benzene from the membrane, and continued for longer than a week. After having finished the first series of measurements, the sample membrane was leached for about a week, and the second series of measurements was performed. Leaching in methanol were performed again between the second and the third series of measurements.

It is seen in Fig. 4.9 that P at 30°C, for example, decreases by about two orders of magnitude over the period of about 100 days under vacuum. The decrease of P occurs in two stages, a rapid decrease in the first stage, from 1 to 15 days, and slow one in the second stage. Whether a further decrease

Figure 4.9 Dependence on total time under vacuum of P for isobutane in poly(TMSP) membrane. $p = 30$ cm Hg. ○ : 30°C, first series; ● : 30°C, second series; ◐ : 30°C, third series; △ : 50°C, first series; □ : 70°C, first series. (Reproduced from Ref. 68. Copyright 1986 The Society of Polymer Science, Japan.)

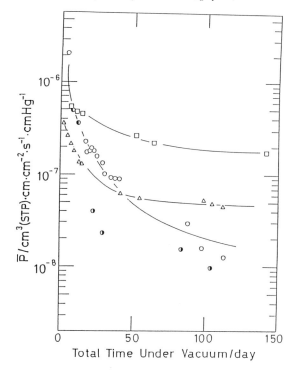

of P after 100 days will occur or not has not yet been confirmed. However, the rate of depression of P, even if a further decrease occurs at longer times, is anticipated to be extremely low. The aging process in the first stage seems to proceed faster at higher temperatures. At 50°C the period of the first stage is shorter than that at 30°C, and the period becomes too short to be observed at 70°C.

In the second series of measurements, the plots coincide almost with those for the first series of measurements. Although the plots for the third series are located somewhat below those for the first and second series, the original high value of P is recovered by the leaching in methanol between the series of permeability measurements. This implies that the microstructure of the poly(TMSP) membrane, which is responsible to the high permeability for the fresh sample and is modified with increasing total time under vacuum, can almost be restored by the leaching procedure. In other words, the process of decreasing P may be considered as a process of purely physical nature.

A similar dependence of P on total time under vacuum as that for isobutane described was observed for permeation of O_2 and N_2 at 30, 50, and 70°C.[67] Also, Nakagawa et al. have observed a similar physical aging process for permeation of various gases through poly(TMSP) membrane.[72,73] Contrary to these observations, no change in either permeability or selectivity for 225 days has been reported.[56]

In connection with the marked depression of P under vacuum, it should be noted that no aging effect on P was observed after a long exposure of the sample membrane to isobutane atmosphere.[71] The rate of depression of P in isobutane atmosphere does not differ substantially from that under vacuum. The result indicates that the physical aging process may be governed by molecular mechanisms that are somewhat different from those of viscoelastic relaxation of polymer solids. Usually the rate of the relaxation process increases with increasing content of low-molecular-weight compound in the polymer.

4.2. Sorption Behavior of Pure Gases and Vapors

Figure 4.10 shows sorption isotherms of CO_2, isobutane, and acetone at 30°C.[74] Henry's law was accurately obeyed for CO_2 at pressures up to 1 atm. As described, the sorption isotherm for isobutane is of the type that can be explained in terms of the dual-mode sorption model. In contrast with the behavior of these simple gases, the shape of the isotherm of acetone is more complex, like those of alcohol vapors.[75]

The characteristic feature of the isotherms of these organic vapors is that all isotherms had two inflection points. The sorption isotherms started out convex to the pressure axis, then became concave and changed to a shape of the type II isotherm of the Brunauer classification[76] at higher pressures. The isotherms were analyzed using a modified form of the equa-

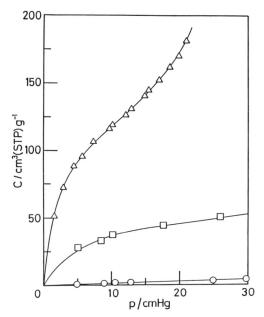

Figure 4.10 Sorption isotherms of gases in poly(TMSP) at 30°C. ○ : CO_2, □ : isobutane, △ : acetone vapor. (Reproduced partly from Ref. 74. Copyright 1988 The Asahi Glass Foundation for Industrial Technology.)

tion derived theoretically by Fowler.[77] By making a proper choice of values of parameters, satisfactory agreement was obtained between calculated and observed isotherms.[75]

4.3. Structures of the Solvent-Cast Membrane

The scanning micrographs for the surface and freeze fracture of the poly(TMSP) membranes used in the permeation and the sorption measurements have indicated the considerable roughness of the surface and the existence of many holes of cylindrical shape inside the membrane.[67,75] The diameters of the holes were about 1 μm. The lengths of the holes were not so long as to penetrate through the membrane, on average several micrometers. The feature of the rough surface and the number and the size of the holes did not change appreciably after prolonged aging under vacuum. Also, an adsorption experiment of N_2 gas at 77.4 K has revealed that the sample membrane has large total surface area, about 550 m^2/g.[75] Furthermore, a relatively low value of bulk density, less than 0.75 g/cm^3, was obtained from measurements of weight and volume (area and thickness) of the membranes.[75]

These observations suggest that the poly(TMSP) membrane has very loose structures compared with membranes of other common glassy polymers, such as polystyrene. The extraordinarily high permeability for various gases and also the characteristic sorption behavior for the organic penetrants of the poly(TMSP) membrane are undoubtedly due to their unique structures. Since the membrane does not show appreciable changes in the rough membrane surface features and in the number and the size of the holes inside the membrane after keeping in vacuum for long time, the physical aging effects on the permeability of gases may be considered to be mostly due to a slow coalescence of intersegmental gaps; that is, the change of microstructure of the membrane on a scale comparable to the size of penetrant molecules seems to be responsible for the observed physical aging process. A relatively wide separation between main chains in the polymer matrix could be formed in the latter stage of the solvent casting by virtue of stiffness of the main chain and bulkiness of the substituents. The molecular motions of the trimethylsilyl and the methyl groups of the polymer will be described in the following.

4.4. Permeation Behavior of Gas Mixtures

In Fig. 4.11, the P values for each component gas in the system of CO_2/isobutane, at a total pressure of 1 atm, and poly(TMSP),[74,78] are plotted by filled symbols against the respective partial pressures in the mixture. Values of P for pure components are also shown in the figure by unfilled symbols. As described before, P for pure CO_2 is independent of p,

Figure 4.11 The dependence of P for CO_2 and acetone vapor on their respective partial pressures at 30°C. The total pressure of the gas mixture is 1 atm. ○ : CO_2 (pure), ● : CO_2 (in mixture), □ : isobutane (pure), ■ : isobutane (in mixture). (Reproduced from Ref. 74. Copyright 1988 The Asahi Glass Foundation for Industrial Technology.)

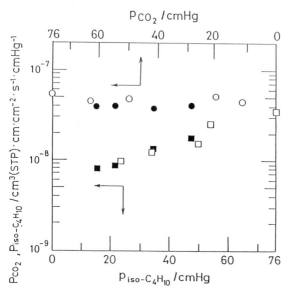

while P for isobutane increases with increasing p. The observed dependence of P on p for both pure components does not conform to the predicted dependence of the dual-mode transport model [Eq. (21)] but is partly in agreement with the predition of the free-volume model [Eq. (26)].

In the permeation of this gas mixture, plots of P for the respective components coincide well with those obtained in pure-component permeation. That is, no effect of one component on another is observed in mixed-gas permeation. This implies that both components permeate through the poly(TMSP) membrane independently.

Values of ln P for both components in mixed-gas permeation have been calculated by Eq. (26). The P versus p relation thus calculated represents well the observed P for isobutane. On the other hand, the calculated relation between P and p for CO_2 is not in agreement with the observed relation.

In the permeation of the mixture of CO_2 and acetone vapor at a total pressure of 1 atm at 30°C, P for CO_2 was influenced a little by the partner component acetone.[74,78] Namely, dissolved acetone molecules enhanced slightly the permeation rate of CO_2. The values of P for CO_2 calculated by Eq. (26) were 1.5 to 2.0 times as high as the observed values in the pressure region studied. On the other hand, the permeation of CO_2 did not interfere with that of aceteone. This may be attributed to the very low solubility of CO_2 in poly(TMSP), which has been shown in Fig. 4.10, as in the case of permeation of CO_2/isobutane mixtures.

As is seen in Fig. 4.12, P for isobutane and that for acetone in permeation of the mixtures are higher than the respective pure-component values.[74,78] The latter values are shown by (thin) solid and dotted lines for isobutane and acetone, respectively. In the figure, the calculated relations, by using Eq. (26), between P and p for isobutane and acetone are represented by (thick) chain and dashed lines, respectively.

The experimental values of P for both components increase with the increase of partial pressure of acetone in the permeating mixture (at total pressure of 1 atm). This may be explained by the high solubility of acetone in poly(TMSP) (Fig. 4.10). The agreement between the experimental and the calculated values of P varies from 25% to a maximum of about 5%. However, by taking into consideration the simplifying assumptions and also the experimental error in the determination of P, we may conclude that the permeation behavior of the isobutane/acetone mixtures, and also most behavior of CO_2/isobutane and CO_2/acetone mixtures, are well explained in terms of the extended free-volume model.

It should be noted here again that all the systems of the gas mixture and poly(TMSP) are not in the rubbery state but in the glassy state at the temperature of experiments. Nevertheless, most of the permeation behavior of the gas mixtures conforms to the prediction of the extended free-volume model, which represents well, as described in Section 1, the permeation behavior of gas mixtures through membranes of rubbery polymers. This

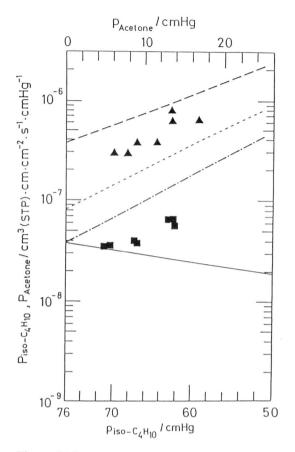

Figure 4.12 The dependence of P for isobutane and acetone vapor on their respective partial pressures at 30°C. The total pressure of the gas mixture is 1 atm. —— : isobutane (pure), ■ : isobutane (in mixture), --- : acetone vapor (pure), ▲ : acetone vapor (in mixture). Calculated values using Eq. (26), – · – : isobutane, – – – : acetone vapor.

suggests that local environments in the poly(TMSP) membrane, which govern the unit jump of the diffusing molecule in the transport processes, would resemble to a considerable extent those in membranes of rubbery polymers. The local environments may result from the microstructures of the poly(TMSP) membrane discussed before and also from the molecular motions of the bulky substituents. In what follows, the latter will be discusssed.

4.5. High-Resolution Solid-State ^{13}C NMR Spectroscopy

Figure 4.13 shows CP (cross-polarization)/DD (dipolar-decoupling) ^{13}C NMR spectra for poly(TMSP) with MAS (magic-angle spinning) at a rate of 6 kHz and without MAS at room temperature.[79] Reso-

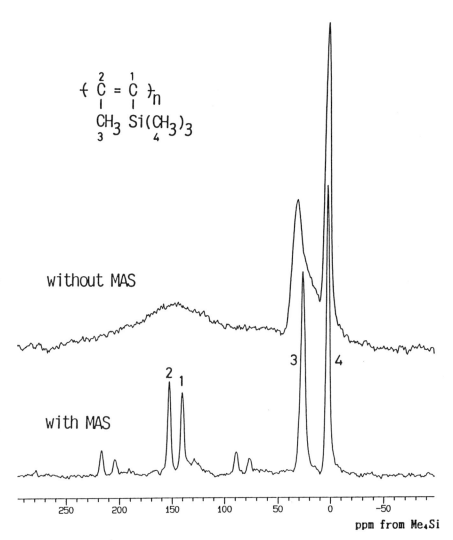

Figure 4.13 CP/DD ^{13}C NMR spectra with MAS at a rate of 6 kHz and without MAS for poly(TMSP) at room temperature.

nance lines for the C_1 and C_2 atoms of the main chain are clearly seen in the spectrum with MAS, but only a very broad peak is seen for these carbons in the spectrum without MAS. This indicates that motions of the main chain are highly restricted, as anticipated from its alternating-double-bond structure. A resonance line for the C_3 carbon of the methyl group, which is attached directly to the main chain, exhibits an upfield shoulder. That is, the axially symmetric spectrum with respect to σ_{11} and σ_{22} axes, which are perpendicular to the C_2–C_3 bond (σ_{33} axis), is observed. On the other hand, the resonance line of the C_4 atom bonded to the Si atom is narrowed by the isotropic motion of the Si–C_4 bond. This means that the motions of three methyl groups are greatly enhanced at room temperature. Very vigorous motion of the Si atom has also been revealed by the CP/DD ^{29}Si NMR spectra with and without MAS at room temperature.[79,80] This motion of the Si atom may enhance the motion of three methyl groups attached to the Si atom.

The ^{13}C NMR powder patterns of the main chain carbon atoms, C_1 and C_2, were measured as envelopes of the spinning sidebands at a rate of 2 kHz.[79] The ^{13}C spectrum is shown in Fig. 4.14. The envelopes for the C_1 and the C_2 atoms are represented in the figure by the dotted and solid lines, respectively. The linewidth of the C_1 atom, to which the Si atom is bonded, is narrower than that of the C_2 atom bonded to the methyl group. This suggests that the mobility of the C_1 atom in the order of 10 kHz is higher than that of the C_2 atom.

The results of NMR studies have thus revealed high mobility of the substituents of poly(TMSP). The characteristic local environments in the solvent-cast poly(TMSP) membrane, which have been considered almost the same as those in membranes of rubbery polymers, can be ascribed partly to the highly mobile nature of the trimethylsilyl group. The loose microstructures discussed previously may be another important factor to explain such local environments that are thought indispensable to develop novel

Figure 4.14 CP/DD ^{13}C NMR spectrum with MAS at a rate of 2 kHz for poly(TMSP) at room temperature.

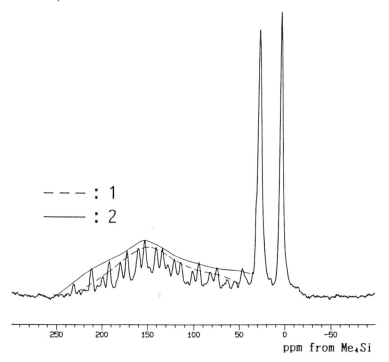

polymeric membranes for gas-separation applications with high productivity.

The motion of the C_1 atom of the main chain is noticeable, although the larger-amplitude motion is strongly restricted at room temperature. In general, the double-bond structure of the main chain is considered to possess a rigid character. This suggests that there would be no substantial difference between the mobilities of these two carbon atoms. However, the observed NMR spectra show clearly that the C_2 atom bonded to the Si atom is higher to some extent than the C_2 atom bonded to the methyl group. This may suggest that the twisting, or fluctuating, motion at 10 kHz frequency is attributable to the double bond in poly(TMSP) at room temperature. This local in-chain movement would be one of the molecular motions occurring in the poly(TMSP) membrane, which results in the depression of gas permeability under vacuum.

REFERENCES

1. For example, C. E. Rogers, in *Polymer Permeability*, J. Comyn, Ed., Elsevier Appl. Sci. Pub., London, 1985, Chap. 2, and the reviews cited in this article.
2. T. Graham, *Phil. Mag.*, *S.4* **32**, 401 (1866).
3. J. Crank, *The Mathematics of Diffusion*, 2nd ed., Clarendon Press, Oxford, 1975.
4. C. E. Rogers, in *Physics and Chemistry of the Organic Solid State*, Vol. 2, D. Fox, M. L. Labes, and A. Weissberger, Eds., John Wiley & Sons, Inc., New York, 1965, Chap. 6.
5. H. Fujita, *Fortschr. Hochpolym.-Forsch.* **3**, 1 (1961).
6. H. Odani, M. Uachikura, Y. Ogino, and M. Kurata, *J. Membrane Sci.* **15**, 193 (1983).
7. H. L. Frisch, *J. Phys. Chem.* **61**, 93 (1957).
8. A. Kishimoto, in *Kobunshi To Suibun (Polymer and Water)*, Soc. Polym. Sci. and Commit. Polym. and Water, Japan Eds., Saiwai Shobo, Tokyo, 1967, Chap. 2.
9. For example, V. Stannett, in *Diffusion in Polymers*, J. Crank and G. S. Park, Eds., Academic Press, London, 1968, Chap. 2.
10. F. P. McCandless, *Ind. Eng. Chem., Proc. Des. Dev.* **11**, 470 (1972).
11. W. J. Koros, *J. Polym. Sci. Polym. Phys. Ed.* **23**, 1611 (1985).
12. W. J. Koros, B. J. Story, S. M. Jordan, K. O'Brien, and G. R. Husk, *Polym. Eng. Sci.* **27**, 603 (1987).
13. S. A. Stern, V. M. Shah, and B. J. Hardy, *J. Polym. Sci. Part B: Polym. Phys.* **25**, 1263 (1987).
14. M. B. Moe, W. J. Koros, and D. R. Paul, *ibid.* **26**, 1931 (1988).
15. S. A. Stern, Y. Mi, H. Yamamoto, and A. S. Clair, *ibid.* **27**, 1887 (1989).
16. S. A. Stern, in *Membrane Separation Processes*, P. Meares, Ed., Elsevier, Amsterdam, 1976, Chap. 8.
17. V. T. Stannett, W. J. Koros, D. R. Paul, H. K. Lonsdale, and R. W. Baker, *Adv. Polym. Sci.* **32**, 69 (1979).
18. H. L. Frisch and S. A. Stern, *CRC Crit. Rev. Solid State Mater. Sci.* **11**, 123 (1983).

19. J. H. Petropoulos, *Adv. Polym. Sci.* **64**, 93 (1985).
20. D. R. Paul and W. J. Koros, *J. Polym. Sci. Polym. Phys. Ed.* **14**, 675 (1976).
21. W. J. Koros, R. T. Chern, V. T. Stannett, and H. B. Hopfenberg, *ibid.* **19**, 1513 (1981).
22. W. J. Koros, *ibid.* **18**, 981 (1980).
23. W. J. Koros, *ibid.* **23**, 1611 (1985).
24. B. J. Story and W. J. Koros, *J. Polym. Sci. Part B: Polym. Phys.* **27**, 1927 (1989).
25. G. R. Mauze and S. A. Stern, *J. Membrane Sci.* **12**, 51 (1982).
26. S. Zhou and S. A. Stern, *J. Polym. Sci. Part B: Polym. Phys.* **27**, 205 (1989).
27. S.-M. Fang, S. A. Stern, and H. L. Frisch, *Chem. Eng. Sci.* **30**, 773 (1975).
28. S. A. Stern, G. R. Mauze, and H. L. Frisch, *J. Polym. Sci. Polym. Phys. Ed.* **21**, 1275 (1983).
29. H. Fujita, A. Kishimoto, and K. Matsumoto, *Trans. Faraday Soc.* **56**, 424 (1960).
30. M. L. Williams, R. F. Landel, and J. D. Ferry, *J. Am. Chem. Soc.* **77**, 3701 (1955).
31. H. Odani, M. Uchikura, and M. Kurata, *ACS Polym. Prepr.* **24**(1), 81 (1983).
32. J. S. Vrentas and J. L. Duda, *J. Polym. Sci. Polym. Phys. Ed.* **15**, 403, 417 (1977).
33. M. H. Cohen and D. Turnbull, *J. Chem. Phys.* **31**, 1164 (1959).
34. R. J. Bearman, *J. Phys. Chem.* **65**, 1961 (1961).
35. P. J. Flory, *Principles of Polymer Chemistry*, Cornell Univ. Press, Ithaca, N.Y., 1953, Chap. 12.
36. F. Bueche, *Physical Properties of Polymers*, Interscience, New York, 1962, Chap. 3.
37. T. Masuda, E. Isobe, T. Higashimura, and K. Takada, *J. Am. Chem. Soc.* **105**, 7473 (1983).
38. Y. Kawakami, H. Kamiya, and Y. Yamashita, *J. Polym. Sci., Polym. Symp.* **74**, 291 (1986).
39. Y. Nagasaki, M. Suda, and T. Tsuruta, *Makromol. Chem., Rapid Commun.* **10**, 255 (1989).
40. T. Masuda and T. Higashimura, *Adv. Polym. Sci.* **81**, 121 (1986).
41. T. Masuda and T. Higashimura, *Acc. Chem. Res.* **17**, 51 (1984).
42. T. Masuda and T. Higashimura, in *Silicon-Based Polymer Science* (Adv. Chem. Ser. No. 244), J. M. Zeigler and F. W. G. Fearon, Eds., Am. Chem. Soc., 1990, Chap. 35.
43. T. Masuda, E. Isobe, and T. Higashimura, *Macromolecules* **18**, 841 (1985).
44. T. Masuda, E. Isobe, T. Hamano, and T. Higashimura, *Macromolecules* **19**, 2448 (1986).
45. J. Fujimori, T. Masuda, and T. Higashimura, *Polym. Bull.* **20**, 1 (1988).
46. E. Isobe, T. Masuda, T. Higashimura, and A. Yamamoto, *J. Polym. Sci. Part A: Polym. Chem.* **24**, 1839 (1986).
47. T. Masuda, E. Isobe, T. Hamano, and T. Higashimura, *J. Polym. Sci. Part A: Polym. Chem.* **25**, 1353 (1987).
48. Air Products and Chemicals, Inc., U.S. Patent 4,759,776 (1988).
49. T. Masuda, T. Matsumoto, T. Yoshimura, and T. Higashimura, *Macromolecules*, **23,** 4902 (1990).

50. Y. Abe, T. Masuda, and T. Higashimura, *J. Polym. Sci. Part A: Polym. Chem.* **27**, 4267 (1989).
51. T. Masuda, T. Hamano, K. Tsuchihara, and T. Higashimura, *Macromolecules* **23**, 1374 (1990).
52. T. Masuda, T. Hamano, T. Higashimura, T. Ueda, and H. Muramatsu, *Macromolecules* **21**, 281 (1988).
53. T. Masuda, B.-Z. Tang, A. Tanaka, and T. Higashimura, *Macromolecules* **19**, 1459 (1986).
54. T. Masuda, B.-Z. Tang, T. Higashimura, and H. Yamaoka, *Macromolecules* **18**, 2369 (1985).
55. M. Langsam and L. M. Robeson, *ACS Polym. Prepr.* **29**(1), 112 (1988).
56. M. Langsam and L. M. Robeson, *Polym. Eng. Sci.* **29**, 44 (1989).
57. T. Masuda, Y. Iguchi, B.-Z. Tang, and T. Higashimura, *Polymer* **29**, 2041 (1988).
58. K. Takada, H. Matsuya, T. Masuda, and T. Higashimura, *J. Appl. Polym. Sci.* **30**, 1605 (1985).
59a. K. Tsuchihara, T. Masuda, and T. Higashimura, *J. Am. Chem. Soc.* **113**, 8548 (1991).
59b. K. Tsuchihara, T. Oshita, T. Masuda, and T. Higashimura, *Polym. J.* **23**, 1273 (1991).
60. T. Higashimura, T. Masuda, and M. Okada, *Polym. Bull.* **10**, 114 (1983).
61. S. Pauly, in *Polymer Handbook*, 3rd ed., J. Brandrup and E. H. Immergut, Eds., J. Wiley, New York, 1989, VI 435.
62. D. W. Van Krevelen, *Properties of Polymers*, 2nd ed., Elsevier, Amsterdam, 1976, pp. 403–25.
63. Y. Ichiraku, S. A. Stern, and T. Nakagawa, *J. Membrane Sci.* **34**, 5 (1987).
64. (a) M. Langsam, M. Anand, and E. J. Karwacki, *Gas Sep. Purific.* **2**, 162 (1988). (b) Air Products and Chemicals, Inc., U.S. Patent 4,657,564 (1987).
65. T. Masuda, *Maku (Membrane)* **13**, 195 (1988).
66. C. E. Rogers, *J. Polym. Sci., Part C* **10**, 93 (1965).
67. H. Shimomura, K. Nakanishi, H. Odani, M. Kurata, T. Masuda, and T. Higashimura, *Kobunshi Ronbunshu* **43**, 747 (1986).
68. H. Kita, T. Sakamoto, K. Tanaka, and K. Okamoto, *Polym. Bull.* **20**, 349 (1988).
69. Y. Nagase, *Koubunshi Kakou (Polymer Processing)* **36**, 268 (1987).
70. J. Crank and G. S. Park, Eds., *Diffusion in Polymers*, Academic Press, London, 1968.
71. H. Shimomura, K. Nakanishi, H. Odani, and M. Kurata, *Rep. Progr. Polym. Phys. Jpn.* **30**, 233 (1987).
72. T. Nakagawa, T. Saito, S. Asakawa, and Y. Saito, *Gas Sep. Purifi.* **2**, 3 (1988).
73. T. Nakagawa, H. Nakano, K. Enomoto, and A. Higuchi, *AIChE Symp. Ser.* **85** (272), 1 (1990).
74. H. Shimomura, T. Uyeda, K. Nakanishi, H. Odani, and M. Kurata, *Asahigarasu Kogyogijutsu Shoreikai Kenkyuhokoku (Rep. Asahi Glass Found. Ind. Technol.)* **53**, 193 (1988).

75. K. Nakanishi, H. Odani, M. Kurata, T. Masuda, and T. Higashimura, *Polym. J.* **19**, 293 (1987).
76. S. Brunauer, *The Adsorption of Gases and Vapors, Vol. I, Physical Adsorption*, Oxford Univ. Press, Oxford, 1945, p. 150.
77. R. H. Fowler, *Proc. Camb. Phil. Soc.* **32**, 144 (1936).
78. T. Uyeda, H. Shimomura, K. Nakanishi, H. Odani, T. Masuda, and T. Higashimura, to be published.
79. T. Uyeda, T. Murata, F. Horii, H. Odani, T. Masuda, and T. Higashimura, to be published.
80. T. Uyeda, K. Takemura, M. Laatikainen, F. Horii, H. Odani, T. Masuda, and T. Higashimura, to be published.

Part III

Separation by Chemical Method

5
Porous Polymer Complexes for Gas Separation

Naoki Toshima and Hiroyuki Asanuma

1. Introduction
2. Design of Adsorbents for Gaseous Molecules
3. Porous Polymers for Gas Separation
4. High Porosity by Complexation with Metal Ions
    4.1. Effect of Washing Solvent
    4.2. Effect of Metal Ion
5. Adsorption of Nitrogen Monoxide by the Porous Dry CR-Fe(II) Complex
    5.1. Preparation of Solid Adsorbent by Immobilization of Metal Complex on the Resin
    5.2. Characterization of the CR–Fe(II)–NO Complex
    5.3. Adsorption Rate
    5.4. Analysis of the Increase in the Surface Area by the Langmuir Equation
    5.5. Mixed-Valence Complex for Effective Adsorbents
    5.6. Desorption of NO and Recyclic Use of the Adsorbent
    5.7. Durability to Oxygen
6. Aqueous Dispersion of CR–Fe(II) Complex as NO Adsorbent
    6.1. Adsorption of NO by the Aqueous Dispersion of the CR–Fe(II) Complex
    6.2. The Equilibrium Constant for the Adsorption of NO by the Dispersion System

6.3. Comparison of the Dry System with Aqueous Dispersion
6.4. Recovery of the Resin
6.5. Simultaneous Adsorption of Nitrogen Monoxide and Sulfur Dioxide
7. Other Inorganic Adsorbents for NO
   7.1. Jarosites
   7.2. FeOOH Systems
8. Concluding Remarks

1. Introduction

Gas separation can be achieved by sorption as well as distillation.[1,2] Sorption can be divided into *adsorption* and *absorption*, which are defined as the sorption of a gas to solid and liquid, respectively.

Gas separation by sorption is largely classified into two categories by purpose:

1. Removal of toxic or unnecessary components (impurity) from the matrix gas.
2. Separation of the mixed gases into each component and concentration of the separated gas (purification).

The properties required for sorbents are different for each purpose. In the former case the absorbate is usually in such low concentration that the sorbent (adsorbent or absorbent) is required to possess a strong affinity for the gas molecule. In the latter case the selectivity for the adsorbate is an indispensable property as well as the affinity. For example, carbon monoxide (CO) is an important basic resource for C_1 chemistry, instead of ethylene in the petroleum chemistry (C_2 chemistry). A large quantity of CO is usually obtained in mixtures with hydrogen, carbon dioxide, and so on by partial oxidation of a natural gas and in the exhaust gas of steelworks. Thus separation, concentration, and purification of CO from the mixture is required for the industrial utilization of CO. In this case the property required for the sorbent is a high selectivity for CO rather than a high affinity. Thus, sorption of CO_2 or H_2O is not preferred. On the contrary, a strong affinity is required for the other case, since CO has another aspect, as a toxic gas for the human body, and has to be removed completely from exhaust gases. In this case the sorbent must have a strong affinity for CO to avoid any discharge of CO to the atmosphere, and the selectivity is less important. The removal of nitrogen monoxide (NO) from exhaust gases is another example. It should be mentioned, however, that adsorption is not the only method for this purpose, since there are many methods to remove industrial gases besides adsorption, for example, oxidation by air. Thus, the properties required of the adsorbent are dependent upon the purpose.

In this chapter the design of polymer adsorbents, especially chemical adsorbents, for gas separation will be discussed, and the importance of porous polymers will be emphasized. As an example of an adsorbent for the concentration and purification of gases, a CO adsorbent supported on porous polystyrene resin will be mentioned briefly, and then a NO adsorbent supported on porous chelate resin will be described.

2. Design of Adsorbents for Gaseous Molecules

For a discussion on the use of adsorption for separation, it is convenient to classify it into (1) physical adsorption and (2) chemical adsorption. The former is mainly based on van deer Waals or electrostatic interactions, so that many porous inorganic materials, such as activated charcoal, silica gel, or alumina, are applicable as adsorbents.[2] The interaction between the adsorbate and the surface, however, is neither so strong nor so specific that selective separation or complete adsorption is still not rather difficult. On the contrary, chemical adsorption based on chemical interactions such as coordination of the sorbate onto the surface of the adsorbent is specific compared with the physical one, so that selective separation is easily achieved. Thus the concentrated recovery or the removal of a trace amount of toxic gas is successfully achieved when the appropriate chemical adsorbent is chosen. Although chemical adsorption has such advantages, however, applicable sorbates are limited because not all of the gases react chemically with the surface.

A chemical adsorbent can be conveniently prepared using metal complexes. Molecules with multiple bond(s) or a nonbonding electron, such as oxygen, carbon monoxide, ethylene, and ammonia, can reversibly form a coordination complex with transition-metal ions. If a *proper metal ion* (with

Figure 5.1 Schematic illustration of the adsorbent composed of a metal complex covalently immobilized on a polymer matrix.

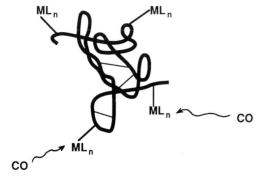

Table 5.1 Metal Complexes Applied to the Sorption of Gaseous Molecules

Gaseous Molecule	Binding Metal	Metal Complex		Polymer Resin as Support	(Ref.)
		Monomer	(Ref.)		
CO	Cu(I)	CuAlCl$_4$	(3–5)	Polystyrene	(6,7)
		Cu(NH$_3$)$_2$	(8)	Polystyrene with amino group	(10)
		Cu(en)^a	(9)		
C$_2$H$_4$	Cu(I)	Cu(en)	(11)	Polystyrene with amino group	(12)
	Ag(I)	AgAlCl$_4$	(13)	Polystyrene	(14)
O$_2$	Co(II)	Co(salen)^b	(15)	POMPy, POMIm^c	(16)
NO	Fe(II)	Fe(EDTA)	(17,18)	Chelate resin with IDA^d group	(19–21)
N$_2$	Mn(I)	C$_5$H$_5$Mn(CO)$_2$	(22)	Polystyrene	(23)

^{a}en: ethylenediamine.
^{g}salen: N,N'-disalicylideneethylenediamine.
^{b}POMPy: poly[(octylmethacrylate)-co-(-vinylpyridine)]; POMIm: poly[(octylmethacrylate)-co-(1-vinylimmidazole)].
^{d}IDA: iminodiacetic acid.

ligands) can be chosen, a solid adsorbent can be successfully prepared by immobilization of the metal ions (or complexes) on a support. There are many methods for the immobilization of metal ions or complexes on supports, and an appropriate way can be chosen in each case. The most popular case is the immobilization of metal ions by ligands, which can be fixed on the polymer chain by covalent bonding, as shown in Fig. 5.1. Table 5.1 shows typical examples of gaseous molecules coordinated to the metal complexes that are applied for immobilization on the polymer. The separation of gases by metal complexes is a promising method due to the specific coordination of gaseous molecules on metal ions.

3. Porous Polymers for Gas Separation

For the separation of gases by metal complexes, the adsorption center is designed by the structure and properties of the metal complexes. The adsorption efficiency of the gas, however, largely depends on the surface area of the support of the adsorbent. In other words, the surface area or porosity of the support can determine the adsorption rate and/or the adsorption capability of the adsorbent (the amount of the adsorbed gas). Thus, high porosity is required for the efficient adsorption of the gas.

In the synthetic polymer resin, especially in the styrene–divinylbenzene copolymer, high porosity was mainly offered by the polymerization technique. When the styrene and divinylbenzene are copolymerized in the absence of a special solvent, transparent and colorless beads (gel-type beads) are obtained. These gel-type beads are homogeneous and do not have physically measurable pores (physical pores). However, a special technique gives the porosity for this copolymer resin. There are two popular methods for the preparation of the porous polymer:

1. Addition of a precipitating solvent as a diluent[24–27]: When a precipitating solvent is added to a polymerization system, a porous polymer is obtained. The solvents used for this purpose have to satisfy the following three conditions:
   a. They are not miscible with water (because the suspension polymerization is always carried out).
   b. They are good solvents for each monomer.
   c. They, however, do not swell the copolymer.

   In the case of the styrene–divinylbenzene system, tert-amyl alcohol, sec-butanol, heptane, and isooctane are usually added and work as precipitants for the copolymer. In the presence of the above additives, the microgel copolymer generated during the polymerization is separated from the system because the added solvent does not have affinity for the copolymer. The polymerization proceeds by forming a three-dimensional network connecting each microgel.[27] The copolymer beads thus obtained consist of many microgels, as illustrated in Fig. 5.2. Thus, physical pores are generated as apertures in the microgels.

2. Addition of a linear polymer as a pore-forming reagent[28,29]: This method is based on the addition of a soluble linear polymer that is incorporated during the copolymerization. After the copolymerization producing beads, the linear polymer is eluted by a solvent. Thus, holes remain in the polymer beads after the elution. The appearance of these polymer beads is similar to that of the gel-type copolymer, since the polymerization mechanism of both polymer beads is almost the same.

These porous polymer beads offer good support for the solid adsorbent. For example, a solid adsorbent of carbon monoxide (CO) with high adsorbing capability can be prepared by immobilizing $CuAlCl_4$ on the porous polystyrene beads.[6] The adsorbent prepared from gel-type copolymer beads instead of the porous ones does not achieve rapid adsorption of CO, as shown in Fig. 5.3. Efficient gas transfer with porous material is also observed in the case of an oxygen–Co(salen) membrane system using hollow fiber as matrix.[30] Details on oxygen and carbon monoxide separation will be described in Chapters 6 and 7, respectively.

Porous polymer beads prepared by the special polymerization technique have physical pores. The network of the polymer is not rigid but is rather flexible, and the porosity varies by treating the polymer with an organic solvent after polymerization. The porosity of the beads of styrene–divinyl-

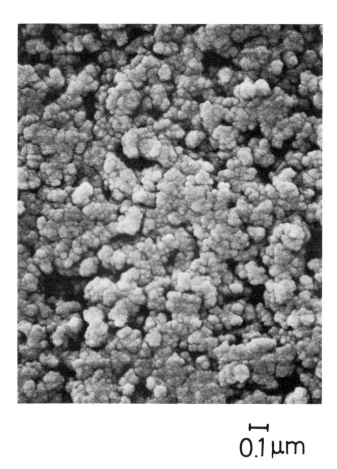

0.1 μm

Figure 5.2 SEM photograph of porous styrene–divinylbenzene copolymer beads.

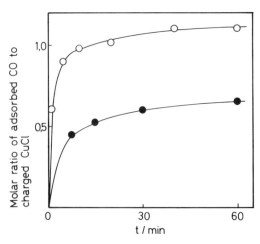

Figure 5.3 Adsorption of CO by the adsorbents prepared from macroporous-type polystyrene (○) and gel-type polystyrene (●).

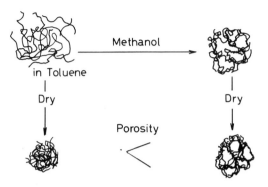

Figure 5.4 Change of porosity by the solvent treatment.

benzene copolymer decreases by desiccating from toluene, which is used as a swelling solvent for the polymer beads. On the contrary, when the beads swollen with toluene are washed with methanol and/or acetone, solvents that cause less swelling, followed by desiccation, the porosity was observed to increase, as illustrated in Fig. 5.4. Haupke and Pientka first reported the above change in the porosity by treatment with organic solvents,[31] and Wieczorek et al. discussed the mechanism of the solvent effect on the porosity.[32,33] These porosity changes are also observed in other polymer beads like acrylic acid resin,[34] styrene sulfonate resin,[35] and chelate resin containing iminodiacetic acid moieties, which we describe in the next section. Porous polymer prepared by this technique can be used for the removal of nitrogen monoxide at low concentration, as is described in the following section.

4. High Porosity by Complexation with Metal Ions

Another technique to control porosity has recently been developed by the present authors, which is based on both the chelation and the solvent treatment. The porosity of beads of the styrene–divinylbenzene copolymer having iminodiacetic acid (IDA) moieties (chelate resin; CR, the chemical structure is shown in Fig. 5.5) can be controlled by the formation of complexes with metals, followed by treatment with organic solvents. The high porosity is successfully achieved by both CR–Fe(III) complexation and desiccation from methanol.[36] The polymer–metal complex beads with high surface area, especially the iron(II)–chelate-resin complex, can be directly applicable to the removal of nitrogen monoxide, since polyamine N-carboxylato-Fe(II) complexes can coordinate to nitrogen monoxide even though its concentration is as low as parts per million. Now the porosity control in the chelate-resin–metal complex system will be described.

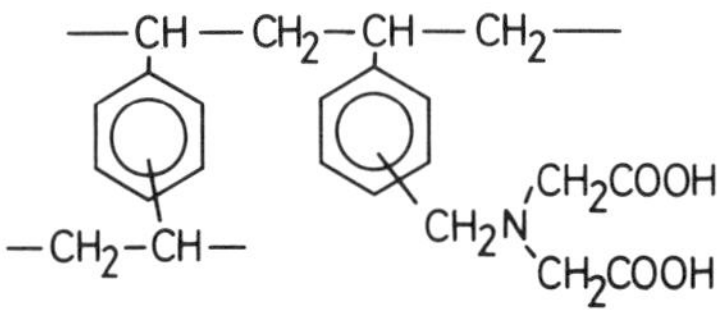

Figure 5.5 Structure of chelate resin containing iminodiacetic acid (IDA) moieties.

4.1. Effect of Washing Solvent

Porosity control in the chelate-resin–metal complex system is achieved by the following two steps: (1) the formation of metal complex with IDA moeities in the chelate resin and (2) desiccation of the water-swollen resin complex having been washed with an organic solvent. The obtained resin complex has an irregular surface, as the SEM photograph in Fig. 5.6 shows.

When the water-swollen CR–Fe(II) complex is dried after having been washed with organic solvents, the porosity (surface area) increases. The porosity is in good agreement with the order of the solubility parameter of the organic solvent, as shown in Table 5.2 and Fig. 5.7.

The increase in the specific surface area upon washing with an organic solvent can be explained as follows. The chelate-resin complex swollen in water is in a rubberlike state. When the solvent (water) in the resin in the rubberlike state is exchanged with the organic solvent miscible with water, many microgels will be generated by a similar phenomenon to the "reprecipitation" of the polymer by addition of a poor solvent to the polymer solution. During the exchange of the solvent, the resin complex is transformed from the rubberlike state into a glassy state because of the poor swelling in the organic solvent.[32,33] In this glassy state, the shrinkage of the resin complex by drying is suppressed by the low mobility of the polymer. Consequently, the surface area increases because the pores produced by drying the resin complex after solvent exchange remain. On the contrary, the resin complex swollen in water shrinks during the evacuation of water in it, since the polymer chains have a high mobility in the rubberlike state. In fact, the resin complex prepared by washing with an organic solvent is lower in apparent density than that prepared by washing with water. This explanation can also be supported by the low porosity of the resin complex freeze dried from water.

The surface area of the chelate-resin–Fe(II) complex increases with an increase in the solubility parameter δ in the order of acetone < ethanol < methanol,[37] as is shown in Table 5.2. The data are plotted in Fig. 5.7. The parameter δ corresponds to the square root of the cohesive energy density $(E/V)^{1/2}$ where E and V are the cohesive energy (energy of vaporization) and volume of unit weight of the solvent, respectively.[37] If the solubility parameters δs of two solvents are close to each other, it indicates that they

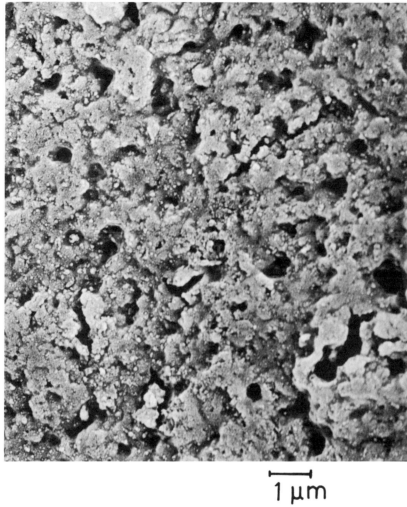

Figure 5.6 SEM photoghraph of the surface of the CR–Al(III) complex.

Table 5.2 Effects of the Washing Organic Solvent on the Specific Surface Area of the Chelate Resin-Immobilized Fe(II) Complex

Solvent	Solubility Parameter[a] $\delta/10^3\ [(J\ m^3)^{1/2}]$	Specific surface area[b] $(m^2\ g^{-1})$
Chloroform	19.0	2.3
(Water	47.9	3.2)
Acetone	20.3	17.1
2-Propanol	23.5	31.7
Ethanol	26.0	31.1
Methanol	29.7	43.1

[a]Quoted from Ref. 37.
[b]Measured by a BET method.

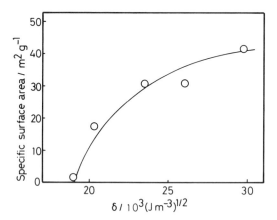

Figure 5.7 Relationship between the specific surface area and the solubility parameter (δ) of the solvent used for washing the water-swollen CR–Fe(II) complex.

have high mutual miscibility.[37] The resin complex prepared from chloroform, whose δ is different from that of water, has a small surface area. On the other hand, the maximum porosity is obtained by the use of methanol, whose δ is close to that of water. These facts indicate that mutual miscibility with water contributes to the surface area (porosity) of the resin complex. Probably a solvent with δ close to that of water, like methanol, will be able to penetrate into the inner portions of the resin complex and be exchanged with water to make pores through "reprecipitation."

4.2. Effect of Metal Ion

The porosity of the resin complex is affected not only by the solvent used for washing but also by the metal ions immobilized on the chelate resin, as shown in Table 5.3. A great increase in the surface area is observed in the case of the chelate resin complexed with high-valent cations. For the CR–Fe(III) complex the specific surface area is about 300 m^2 g^{-1}, which is larger by 4 orders of magnitude than that of the original CR–Na(I) resin dried from water.

The specific surface area also depends on the amount of metal ion immobilized on the chelate resin, as shown in Fig. 5.8. The surface area dramatically increases when the ratio (R) of the immobilized Fe(II) ion to IDA moieties in the chelate resin is larger than 0.25. However, almost no increase in the surface area is observed below this point.

The pores obtained by this method is very small, as Fig. 5.9 shows. In the case of the CR–Fe(III) system, the small pores with radius (r) below 2.5 nm are mainly observed. A similar trend was also observed with CR–Fe(II) desiccated from methanol. However, the CR–Fe(III) system desiccated from water, whose surface area was as small as 5.9 m^2 g^{-1}, had few pores

Table 5.3 Apparent Volume and Specific Surface Area of Various Chelate Resin-Immobilized Metal Complex

Metal	Water[a]		Methanol[a]		Molar Ratio of Immobilized Metal Ion to IDA Moieties
	Volume[b] (cm³ g⁻¹)	Surface Area[c] (m² g⁻¹)	Volume[b] (cm³ g⁻¹)	Surface Area[c] (m² g⁻¹)	
Li(I)	1.21	1.1	1.47	6.8	2.0
Na(I)	1.28	<0.1	1.50	3.3	2.0
K(I)	1.26	<0.1	1.39	1.6	2.0
Mg(II)	1.31	2.5	1.45	22.6	1.0
Ca(II)	1.34	3.6	1.48	33.3	0.94
Sr(II)	1.32	3.2	1.43	25.1	1.0
Ba(II)	1.33	3.5	1.42	19.0	1.0
Al(III)	1.33	4.7	1.53	54.1	0.77
Fe(III)	1.68	5.9	2.13	329.2	1.0

[a] Solvent used for washing the resin-immobilized metal complex.
[b] Volume per 1 g of the chelate resin in the sodium form.
[c] BET surface area per 1 g of the chelate resin in the sodium form.

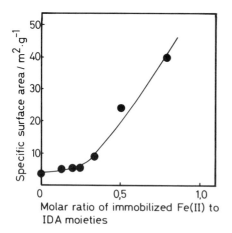

Figure 5.8 Dependence of the specific surface area on the molar ratio of immobilized Fe(II) ions to IDA moieties.

with $r < 10$ nm. Thus it can be said that the pores generated by this method are as small as molecular level.

Metal cations with high valency are immobilized on the chelate resin through complex formation with IDA moieties, producing the 1:1 complex shown in Fig. 5.10. In fact, the amount of trivalent cations immobilized on the resin is more than that required for compensating the anion in the resin ($R = 0.66$), as shown in the last column of Table 5.3. Thus, the CR–trivalent-cation complexes, CR–Al(III) and CR–Fe(III), acquire anion exchange ability derived from the residual positive charge of $M(III)$–IDA complex, as shown in Table 5.4. This is not true for the Al(III)–polyacrylic-acid resin [WK–Al(III)] complex. The suppression of the shrinkage by washing with an organic solvent, which is discussed in the previous section,

Figure 5.9 Pore size distributions in CR–Fe(III) complex prepared by use of methanol (——) and water (···), and in CR–Fe(II) (– - –) prepared by use of methanol. Δv is the sum of the total volume for the pores having radii between r and $r + \Delta r$.

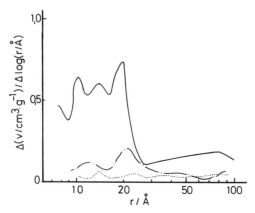

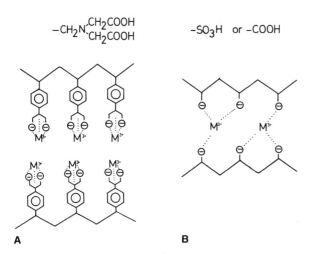

Figure 5.10 Schematic illustration for the structure of the trivalent-metal-ion–resin complex with various moieties. A: Chelate resin-trivalent-cation complex. A metal ion is immobilized on each IDA moiety through coordination. B: Sulfonate (or Carboxylate) resin-trivalent-cation complex. Ionic moieties are crosslinked by trivalent cations through electrostatic interaction.

also occurs in the case of CR–high-valent-cation complex. In the case of the CR complex, however, the chelating structure shown in Fig. 5.10(a) is additionally considered to contribute to the high porosity. Especially in the case of the trivalent metal cation is the electrostatic repulsion between the residual positive charges considered to prevent shrinkage of the resin complex, as illustrated in Fig. 5.10(A). Moreover, the chelating structure will probably promote the transition of the rubberlike state into the glassy state by washing with an organic solvent. On the contrary, other resins having carboxylate or sulfonate groups, which immobilize the metal ions through electrostatic interaction and cannot form the chelating structure, do not exhibit an increase as in the chelate resin. Cations with higher valency decrease the surface area of sulfonate or acrylate resin rather than increase it, as Table 5.5 shows. This is explained by the high valent cation crosslinking each monomer unit through electrostatic interaction, which is illus-

Table 5.4 Anion Exchange Ability of Various Resin–Metal-Ion Complexes

Resin Complex[a]	Anion Exchange Ability[b]
CR–Na(I)	No
CR–Mg(II)	No
CR–Al(III)	Yes[c]
WK–Al(III)	No

[a]CR: chelate resin containing IDA moieties; WK: polyacrylate resin.
[b]Anion exchange ability was examined by use of aqueous solution of $PtCl_6^{2-}$.
[c]Exchanged amount of $PtCl_6^{2-}$ was almost the same amount as that of the residual positive charge in the CR–Al(III) complex.

Table 5.5 Surface Area of Various Resin-Immobilized Metal Complexes

Resin	Metal	Surface Area	Metal/Moieties[a]
Polystyrene	Na(I)	0.2	1.0
sulfonate	Mg(II)	<0.1	0.49
resin	Al(III)	<0.1	0.29
Polyacrylate	Na(I)	0.39	1.0
resin	Mg(II)	<0.1	0.42
	Al(III)	<0.1	0.04

[a]Molar ratio of immobilized metal ions to the sulfonate or carboxylate moieties in the resin.

trated in Fig. 5.10(B). This figure also explains the lack of the anion exchange ability of the Al(III)–polyacrylic-acid resin complex, since this conformation does not maintain the residual positive charge.

5. Adsorption of Nitrogen Monoxide by the Porous Dry CR–Fe(II) Complex

The concentration of nitrogen monoxide (NO) in the atmosphere has not decreased but rather increased in the past decade in spite of much effort.[38] Since NO is toxic to the human body and causes photochemical smog through nitrogen dioxide, advanced technology for the removal of NO is strongly desired. There are a lot of methods to remove NO, for instance, catalytic reduction of NO into innoxious nitrogen with ammonia or hydrogen, absorption of NO with absorbents, and so on.[39–44] Among these methods, the complex of Fe(II) with polyamine N-carboxylate [ethylenediaminetetraacetate (EDTA), nitrilotriacetate (NTA), and IDA] is one of the promising ways because of its high capability for absorbing NO through complex formation.[45–47]

Solid adsorbent can be prepared by immobilization of the ligand on a polymer chain, as we have mentioned. For this purpose the chelate resin involving IDA moieties, which we have discussed in the previous section, is available as the material. The porous polymer beads of the chelate-resin–Fe(II) complex are quite useful as an adsorbent for nitrogen monoxide,[19–21] and they can be used for separation of NO. This section will describe the preparation and the characterization of the porous dry beads of CR–Fe(II) complex and their adsorption properties of nitrogen monoxide.

5.1. Preparation of Solid Adsorbent by Immobilization of Metal Complex on the Resin

The adsorbing property of an adsorbent is strongly dependent on its surface area. Thus, the above method to increase the porosity of the polymer complex must be effective to achieve the high adsorption capability of the polymer complex adsorbent.

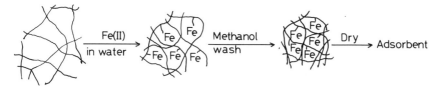

Figure 5.11 Preparation of CR–Fe(II) complex with a high surface area for the adsorption of NO.

The dry polymer beads of CR–Fe(II) complex can be prepared according to the method shown in Fig. 5.11.[19] Thus, the commercial chelate resin in the sodium form is added to the aqueous solution of Fe(II) sulfate or chloride under nitrogen atmosphere to avoid the oxidation of Fe(II) to Fe(III). The obtained mixtures are shaken to achieve complete immobilization of Fe(II) ions on the resin. The white resin exhibits a slightly greenish color due to the complexation. The supernatant of the resulting mixtures is removed by decanting, and the solid part is dried after washing with methanol in order to offer a high surface area. The CR–Fe(II) complex thus obtained shows so high a capability that almost all the NO is adsorbed from 6 dm^3 of nitrogen containing 1000 ppm of NO within 25 min, as illustrated by the closed circles in Fig. 5.12. By contrast, the CR–Fe(II) complex prepared by washing with water ($S = 3.2$ m^2 g^{-1}) does not exhibit such a high adsorbing capability, as illustrated by the open circles in Fig. 5.12.

Figure 5.12 Adsorption curves for NO by the CR–Fe(II) complex prepared by drying after washing with methanol (●) and water (○). These adsorbents were prepared from 21.4 g of chelate resin and 8.72 g (31.4 mmol) of $FeSO_4 \cdot 7H_2O$. The adsorption experiment was carried out by circulating 6 dm^3 of nitrogen containing 1000 ppm of NO at room temperature at the rate of 1.6 dm^3 min^{-1}.

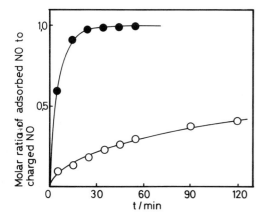

5.2. Characterization of the CR–Fe(II)–NO Complex

The dry beads of CR–Fe(II) complex exhibit a light green color with a maximum peak in the electronic spectrum at around 300 nm. The color of the CR–Fe(II) complex changes from light green to brown at contact with the NO gas, resulting in new shoulder peaks at around 470 and 610 nm in the electronic spectrum (Fig. 5.13),[20] which correspond to the charge-transfer bands between NO and Fe(II) ion.[17,18]

The ESR spectrum also gives information about the coordinative adsorption of NO on the CR–Fe(II) complex. When the CR–Fe(II) complex is exposed to NO, an asymmetric, strong, and broad signal with unresolved shoulders at around $g = 2.03$ appears, as is shown in Fig. 5.14,[20] which is characteristic to the Fe(II)–NO complexes derived from an NO molecule with a free electron.[48] The appearance of the signal at around $g = 2.03$, not in the lower magnetic field, indicates the formation of a low-spin ($S = 1/2$) Fe(II)–NO complex. When the ligand field is strong enough, a low-spin complex is known to be formed. In the present system, since the IDA moieties in the chelate resin work as strong ligands, the low-spin complexes of Fe(II)–NO are produced. Thus, the CR–Fe(II) adsorbs NO through complex formation of NO on Fe(II) as an aqueous solution of Fe(II)–EDTA does. The reaction scheme is shown in Fig. 5.15.

5.3. Adsorption Rate

The adsorbing capability for NO, especially the adsorption rate of the dry adsorbent prepared by washing, coincides well with the specific surface area of the adsorbent. Figure 5.16 shows plots of the adsorption rate of NO against the specific surface area. The adsorption rate linearly increases with an increase in the surface area up to 24 $m^2 g^{-1}$. This fact clearly demonstrates that the increase in the surface area contributes directly to an increase in the adsorption rate. In other words, the porosity of the polymer beads is an important factor for a good polymer complex adsorbent. Over a surface area of 24 $m^2 g^{-1}$, the adsorption rate is so fast that the adsorption apparatus used in this experiment cannot follow the increase in the adsorption rate, and so remains constant.

5.4. Analysis of the Increase in the Surface Area by the Langmuir Equation

The adsorption of NO by the dry CR–Fe(II) complex is a reversible coordination of NO on Fe(II) ions in the complex. The coordination number of NO to Fe(II) ions is one or two in the aqueous solution of a Fe(II)–polyamine N-carboxylate complex system.[18,45] In the case of the CR–Fe(II) system, the adsorption of NO on Fe(II) proceeds by 1:1 complex formation, which can be concluded by the Langmuir adsorption isotherm.

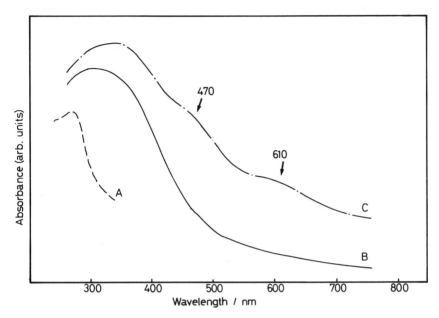

Figure 5.13 Electronic spectra of the NO adsorbent and related materials. A (– –) : chelate resin, B (———) : CR–Fe(II), and C (–·–) : CR–Fe(II)–NO.

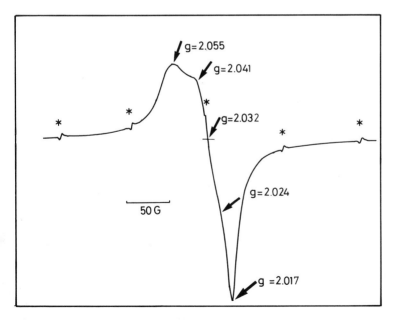

Figure 5.14 ESR spectrum of CR–Fe(II)–NO complex at a central magnetic field. Five weak signals with asterisks are derived from a Mn^{2+} marker.

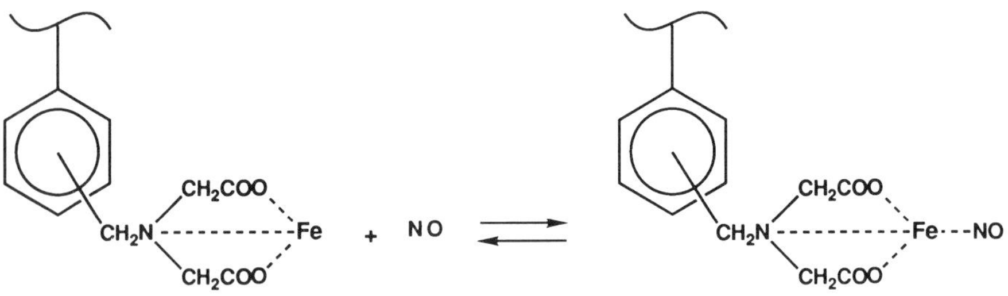

Figure 5.15 Coordination of NO to the CR–Fe(II) complex.

From the Langmuir adsorption isotherm, the amount of effective Fe(II) ions and the equilibrium constant (K/atm^{-1}) are also calculated.[49] When the adsorption of NO by the CR–Fe(II) complex proceeds according to the Eq. (1), Eq. (2) can be derived from Eq. (1);

$$CR\text{–}Fe(II) + NO \underset{}{\overset{K}{\rightleftharpoons}} CR\text{–}Fe(II)\text{–}NO \quad (1)$$

$$1/K = (P/[CR\text{–}Fe(II)\text{–}NO])\,[CR\text{–}Fe(II)] - P, \quad (2)$$

where P is the partial pressure of NO (atm) and [CR—Fe(II)–NO] is the amount of the NO adsorbed at equilibrium (mmol). From the linear plots of P against P/[CR–Fe(II)—NO], the effective Fe(II) and the dissociation constant $(1/K)$ can be calculated from the slope and the intercept of the ordinate, respectively.

Figure 5.17 illustrates the resulting linear plots of P and P/[CR–Fe(II)–NO] by the dry CR–Fe(II) complexes prepared by washing with methanol, ethanol, and acetone from the adsorption isotherm. As Fig. 5.17 shows, the Langmuir plot of each adsorbent forms a straight line, demonstrating that

Figure 5.16 Correlation of the specific surface area with the adsorption rate of NO by CR–Fe(II) complexes prepared by the use of various organic solvents.

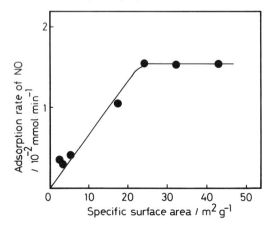

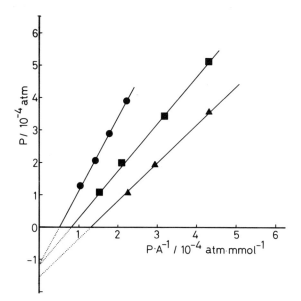

Figure 5.17 Langmuir plots of the adsorbent prepared by drying after the swollen resin complex has been washed with methanol (●), ethanol (■), and acetone (▲).

the coordination of NO proceeds through the 1:1 complex formation of NO to the Fe(II) ion. The K value of each CR–Fe(II) complex is around 8×10^3 atm^{-1}, which does not depend on the solvents used for washing (Table 5.6). The equilibrium constant obtained is even larger than that of an aqueous solution of the Fe(II)–EDTA complex (5.6×10^3 atm^{-1}).[17]

On the contrary, the amount of effective Fe(II) ions increases with an increase in the surface area caused by washing with the organic solvent. Thus, it is clarified that washing with organic solvent does not raise the complex formation constant K, but raises the amount of the effective Fe(II) ions through an increase in the surface area caused by washing with the organic solvent. In fact, the amount of effective Fe(II) ions is proportional to the surface area, as shown in Fig. 5.18.

Table 5.6 Effect of the Washing Solvent on the Effective Amount of Fe(II) Ions and the Complex Formation Constant (K) at Room Temperature

Solvent	Effective Fe(II) (mmol)	Surface Area (m^2 g^{-1})	K (10^3 atm^{-1})	$1/K$ (10^{-4} atm)
Methanol	2.27(0.282)[a]	43.1	8.33	1.20
Ethanol	1.47(0.197)	31.1	8.33	1.20
Acetone	1.16(0.144)	17.1	6.67	1.50

[a]Effective amount of Fe(II) ions per 1 g of the dry-type CR–Fe(II) complex.

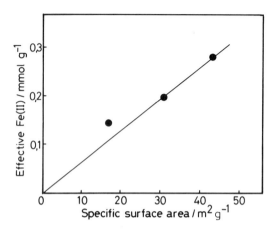

Figure 5.18 Relationship between the specific surface area of each adsorbent and the effective amount of Fe(II) ions calculated from the slope of Fig. 5.17.

5.5. Mixed-Valence Complex for Effective Adsorbents

In the previous paragraph, it has been demonstrated that high surface area is required for high adsorbing capability. However, the CR–Fe(II) complexes do not completely exhibit the potential adsorbing capability, since the surface area is not sufficiently large. Thus, Fe(III) is intended to be incorporated into the CR–Fe(II) complex in order to make the complex porous,[36] as mentioned in Section 4.

The chelate-resin–mixed-valence-iron complex coordinates NO so fast that almost all (more than 99%) of the NO in the system was removed by adsorption within 35 min, as is shown in Fig. 5.19. On the other hand, fast adsorption was not achieved by the adsorbent composed of the chelate resin and 5.0 mmol of Fe(II) ions. The specific surface area of this Fe(II) complex was 5.3 $m^2\ g^{-1}$, which was much less than that of the mixed-valence-iron complex (128.0 $m^2\ g^{-1}$), composed of 10.2 mmol of Fe(III) ions and 3.56 mmol of Fe(II) ions. This fact indicates that, since Fe(III) ions are not active in the coordination of NO, the rapid coordination achieved in the case of the chelate-resin–mixed-valence-iron complex is attributed to the high surface area caused by the Fe(III) ions simultaneously introduced into the resin, which has already been discussed.

Thus, each metal ion plays a different role; Fe(II) ions work as active centers for the adsorption of NO, and Fe(III) ions make the CR complex porous (Fig. 5.20). The Langmuir plots in Fig. 5.21 clearly exhibit the cooperative function of each metal ion. The dissociation constant of this mixed-valence complex (1.0×10^{-4} atm) is the same as that of CR–Fe(II) complex (1.2×10^{-4} atm), indicating that the Fe(II) ions in the mixed-valence complex function as the adsorption site.

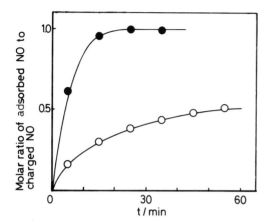

Figure 5.19 Adsorption curves of NO by the chelate-resin–mixed-valence-iron complex (●) involving 10.2 mmol of Fe(III) ion and 3.56 mmol of Fe(II) ion with 128.0 m^2 g^{-1} of specific surface area, and the chelate-resin–immobilized-Fe(II) complex (○) involving 5.0 mmol of Fe(II) ion with 5.3 m^2 g^{-1} of the surface area.

The amount of the effective Fe(II) calculated from the slope of the plots in the case of the chelate-resin–mixed-valence-iron complex and the Fe(II) complex are 0.93 and 2.27 mmol, respectively, as listed in Table 5.7. The value of the efficiency R of the Fe(II) ions can be estimated from the ratio of the amount of effective Fe(II) to the amount of immobilized Fe(II). The efficiency R of the chelate-resin–mixed-valence-iron complex is larger than that in the Fe(II) complex due to an increase in surface area, demonstrating that the Fe(III) ions function to make the resin porous.

Thus it can be concluded that the coordination rate or capability of the chelate-resin–immobilized Fe(II) complexes can increase by simultaneously

Figure 5.20 Adsorption of NO with mixed-valence complex. Fe(III) ions make the resin porous, and Fe(II) ions work as the active center.

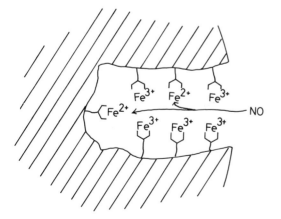

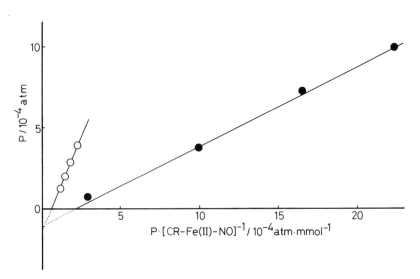

Figure 5.21 Langmuir plots of the chelate-resin–mixed-valence-iron complex (●) composed of 10.2 mmol of Fe(III) ion and 3.56 mmol of Fe(II) ion, and the chelate-resin–immobilized-Fe(II) complex (○) involving 16.2 mmol of Fe(II) ion.

introducing the Fe(III) ions into the resin due to an increase in surface area. The rest of the Fe(II) ions, which have no relation to the coordination of NO, probably exist inside the microsphere, into which the NO molecule cannot go. A further increase in the surface area will be expected to raise the efficiency R.

5.6. Desorption of NO and Recyclic Use of the Adsorbent

The adsorption of NO by CR–Fe(II) is reversible, as mentioned, so that the adsorbed NO can be released by heat treatment. Thus, adsorption–desorption cycles are successfully repeated after the desorption of NO without any significant deterioration of the adsorbent, the data of which are shown in Fig. 5.22.

Table 5.7 Amount of Effective Fe(II) Ion from Langmuir's Plots

Complex	Immobilized Ion		Surface Area ($m^2\ g^{-1}$)	Effective Fe(II) (mmol)	Efficiency R^a
	Fe(III) (mmol)	Fe(II) (mmol)			
CR–Fe(II)	0	16.2	43.1	2.27	0.14
CR–Fe(II), Fe(III)	10.2	3.56	128.0	0.93	0.26

[a]The efficiency R was determined as the ratio of effective Fe(II) calculated from the slope of Fig. 5.21 to the amount of immobilized Fe(II).

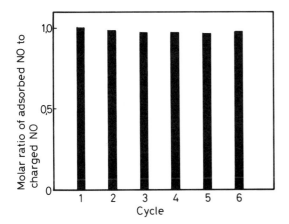

Figure 5.22 Repeated use of the dry-type adsorbent after desorption of NO at 90°C under 3 Torr for 3 h (runs 2–5). Before the sixth run, the NO was released at 100°C under 3 Torr for 6.5 h.

5.7. Durability to Oxygen

For practical use of the sorbent (absorbent or adsorbent), a certain durability to oxygen is important. In the case of the aqueous solution of Fe(II)–EDTA complex as a common absorbent for NO, the solution easily deteriorates with oxygen contaminant through the oxidation of Fe(II) to Fe(III) ions, which are inactive for NO coordination.[46,50]

In contrast, the present CR–Fe(II) dry system has some durability to oxygen in comparison with the aqueous system of Fe(II)–polyamine N-carboxylate complex. Both Fe(II) complexes of CR and IDA can remove more than 90% of NO in a closed circulation system before contact with air. In the case of the aqueous solution of Fe(II)–IDA complex, however, the amount of absorbed NO decreases below 20% after contact with air. The corresponding decrease in Fe(II) ions in the solution is accompanied by a decrease in NO adsorption, as shown in Fig. 5.23 and Table 5.8. In contrast, the decrease in the amount of adsorbed NO with CR–Fe(II) complex is scarcely observed, although half of the Fe(II) ions immobilized are oxidized by the same treatment with air.

In the case of the Fe(II)–IDA solution, the oxidation of Fe(II) ions proceeds with contact to air (Fig. 5.24), and the Fe(II) ions remaining in the solution probably decrease exponentially with time. Consequently, the amount of Fe(II) ions remaining in the solution is 13% of the initially charged Fe(II) ions after contact with air for 60 min, as shown in Fig. 5.24. When the dry-type CR–Fe(II) complex is exposed to air, the oxidation of Fe(II) ions in the complex proceeded for 20 min, as is observed with the Fe(II)–IDA solution. However, further oxidation is suppressed, and the ratio of the oxidized Fe(II) ions to the original ones is 51.9% after contact with air for 60 min.

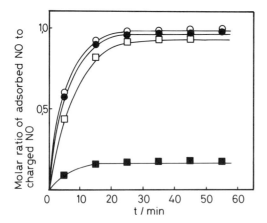

Figure 5.23 Adsorption of NO by the ethanol-washed CR–Fe(II) complex (circle) and the aqueous solution of the Fe(II)–IDA complex (square) before (open) and after (closed) contact with air previously dried by passing through calcium chloride for 1 h.

Table 5.8 Oxidation of Fe(II) Ions by Passing Air

Complex	Treatment	Adsorbed NO[a] (%)	Fe(II)/Fe
Fe(II)–IDA(aq)	none	93.0	1.0
Fe(II)–IDA(aq)	air[b]	18.0	0.131
Fe(II)–CR	none	>99	1.0
Fe(II)–CR	air[b]	89.3	0.519

[a]Initially charged amount of NO is 0.246 mmol (6 dm^3 of nitrogen containing 1000 ppm of NO).
[b]Contacted with dried air at the rate of 1.6 dm^3 min^{-1} for 1 h.

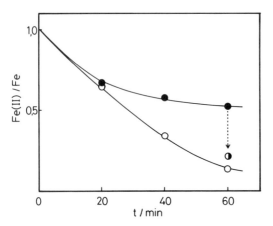

Figure 5.24 Change of the ratio of Fe(II) ions to charged Fe [Fe(II)/Fe] in an aqueous solution of Fe(II)–IDA complex (○) and ethanol-washed CR–Fe(II) complex (●) with time during contact with air previously dried by passing through calcium chloride. The half-filled circle shows the Fe(II)/Fe ratio of the CR–Fe(II) complex after contact with air saturated with water vapor for 60 min.

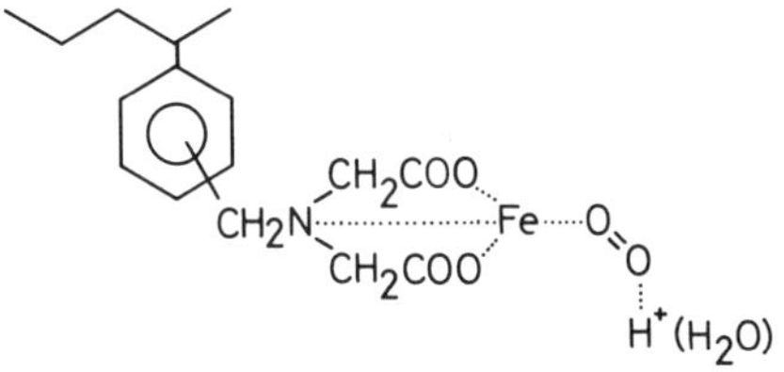

Figure 5.25 A possible mechanism for the oxidation of Fe(II) to Fe(III) by oxygen.

The dry-type CR–Fe(II) complex has distinct durability to oxygen compared with the corresponding aqueous dispersion of the CR–Fe(II), which will be described in the next section. This durability is attributed to the absence of water. In fact, when the dry-type CR–Fe(II) complex is in contact with air saturated with water vapor instead of dry air, the amount of Fe(II) ions remaining in the resin after air treatment decrease to 21.7%, as shown with the half-filled circle in Fig. 5.24. The oxidation of Fe(II) is known to be accelerated by the presence of water according to the reaction equation[51]

$$Fe(II) + 1/4O_2 + 1/2H_2O \rightarrow Fe(III) + OH^-. \tag{3}$$

Thus, when Fe(II) ion, oxygen, and proton (H_2O) are close together, the Fe(II) ion will be oxidized by irreversible electron transfer through the dioxygen complex, as schematically illustrated in Fig. 5.25. These considerations suggest that the oxidation reaction can be suppressed in the absence of water molecules. The dry-type CR–Fe(II) complex has a high enough adsorbing capability to adsorb almost all the NO in the system, although nearly half of the Fe(II) ions are oxidized by contact with air for 60 min.

6. Aqueous Dispersion of CR–Fe(II) Complex as NO Adsorbent

In the previous section, the application of porous dry CR–Fe(II) for the adsorption of NO has been demonstrated. However, a wet system such as an aqueous solution of Fe(II)–EDTA is also promising for the removal of NO because the aqueous solution can remove other pollutants such as sulfur dioxide and dusts involved in an actual exhaust gas. Thus, the CR–Fe(II) complex dispersed in water was also examined for the removal of NO, since this system has advantages in the easy recovery and repeated use of the metal complex compared with the Fe(II)–EDTA system.

6.1. Adsorption of NO by the Aqueous Dispersion of the CR–Fe(II) Complex

Figure 5.26 depicts various NO adsorption curves by aqueous dispersion of chelate-resin-immobilized Fe(II) complexes from nitrogen gas containing 1000 ppm NO, which demonstrates that the CR–

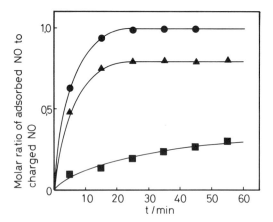

Figure 5.26 Adsorption curves for NO by the aqueous dispersion of the chelate-resin-immobilized Fe(II) complexes prepared from coarse (■) and fine (▲) beads, and the dry-type CR–Fe(II) complex (●) prepared by use of methanol as the solvent for washing.

Fe(II) system also has an adsorbing capability for NO even in the dispersed state in solutions.[52–54] In order to effect rapid adsorption of NO by aqueous dispersion, the CR particles should be required to be small and have high surface area.

Nitrogen monoxide is easily desorbed by treating the NO-adsorbed dispersions with heat, since the equilibrated amount of adsorbed NO decreases with increasing temperature (Table 5.9). The adsorption–desorption cycles can result in the separation and the concentrated recovery of NO.[53,54]

In order to acquire a high adsorbing capability, the carboxyl group is necessary. The NO is scarcely adsorbed by the polystyrenesulfonate resin, on which the Fe(II) ions are immobilized electrostatically (Fig. 5.27). On the contrary, the polyacrylic acid–Fe(II) complexes adsorb 51% of the NO within 60 min. Thus, the complex formation of Fe(II) ions with the carboxyl groups is necessary to achieve the NO adsorption property. In fact,

Table 5.9 Adsorption of NO by the Chelate Resin-Immobilized Fe(II) Complex at Various Temperatures

Temp. (°C)	Adsorbed NO^a (mmol)	Apparent Equilibrium Constant	
		K_cH^{-1} (atm^{-1})	K_c (dm^3 mol^{-1})
25.0	0.191 (0.78)^b	60.1	31.1 × 10^3
39.5	0.134 (0.58)	23.2	14.8 × 10^3
59.8	0.069 (0.29)	7.00	5.35 × 10^3
80.0	0.025 (0.11)	2.02	1.67 × 10^3

^{a}Initial amount of NO = 0.246 mmol.
^{b}Molar ratio of adsorbed NO to charged NO.

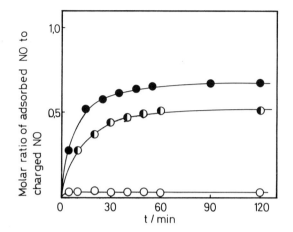

Figure 5.27 Adsorption of NO by the aqueous dispersions of Fe(II) complex immobilized on the chelate resin (●), polyacrylate resin (◐), and polystyrene sulfonate resin (○) from 6 dm³ of the nitrogen gas containing 1000 ppm of NO.

$Fe(II)SO_4$, which is corresponding to the sulfonic acid resin-immobilized Fe(II) complex, scarcely absorbs NO,[54] while acetato-Fe(II)[55] or iminodiacetato–Fe(II) complex does.

In the aqueous dispersion of CR–Fe(II) complex, the active part for the adsorption of NO is the solid part. The color of this complex changes from light green to dark green in contact with NO, resulting in two shoulder peaks at around 450 and 600 nm, as observed in the dry CR–Fe(II) complex system (Fig. 5.13). These two peaks disappear after the desorption of NO. Thus, the CR–Fe(II) complex of this aqueous dispersion system can reversibly adsorb NO by coordinating NO on the Fe(II) ion.

6.2. The Equilibrium Constant for the Adsorption of NO by the Dispersion System

In contrast to the dry CR–Fe(II) system, the above dispersion system is considered to start the adsorption from the dissolution of gaseous NO into water according to

$$NO_{gas} \leftrightharpoons NO_{aq} \quad (4)$$

$n(NO_{aq}) + CR\text{–}Fe(II) \leftrightharpoons CR\text{–}Fe(II) - (NO)_n$. When $n = 1$, (5)

$$K = [CR\text{–}Fe(II)\text{–}NO]/\{[CR\text{–}Fe(II)] [NO_{aq}]\}$$
$$= [CR\text{–}Fe(II)\text{–}NO] \, H/\{[CR\text{–}Fe(II)] \, p_{NO}\} \quad (6)$$
$$K = m_{NO} \, H/\{[CR\text{–}Fe(II)]_0 \, p_{NO}\}$$
$$K \, p_{NO}/H = m_{NO}/[CR\text{–}Fe(II)]_0. \quad (7)$$

Assuming a first-order reaction for the NO adsorption, or a 1:1 complex formation of NO with the Fe(II) ion [$n = 1$ in Eq. (5)], Eq. (6) can be obtained from Eq. (5), where p_{NO} and H are the partial pressure of NO (atm) and Henry's constant, respectively. The amount of Fe(II) ions immobilized on the chelate resin, [CR–Fe(II)] in mmol, can have a linear relation with the amount of adsorbed NO, m_{NO}. Thus, the uncoordinated amount is obtained by subtracting the m_{NO} from the amount of the initially immobilized Fe(II) ions, [CR–Fe(II)]$_0$. As the CR–Fe(II) exists in large excess to the adsorbed NO ([CR–Fe(II)]/m_{NO} = 22 in the present case), however, the amount of unreacted CR–Fe(II), [CR–Fe(II)], can be regarded as the amount of the initially immobilized Fe(II) ions, [CR–Fe(II)]$_0$. Thus, Eq. (6) is transformed into Eq. (7), which indicates that the partial pressure of NO at equilibrium should be in proportion to the amount of NO adsorbed when the 1:1 complex formation proceeds. In fact, the amount of NO adsorbed is in good proportion to the partial pressure of NO,[54] which also indicates that the equilibrium constant K of the adsorbing reaction can be calculated according to Eq. (6).

Table 5.9 shows the equilibrium constant at each temperature obtained according to Eq. (6). The equilibrium constant decreases with an increase in the temperature because of the fall in the amount of NO adsorbed. The equilibrium constants K_c can be expressed as follows by using the temperature T as a parameter from the van't Hoff plots:

$$K_c = (3.32 \times 10^{-4}) \exp(45.6 \times 10^3/RT), \tag{8}$$

where $R = 8.314$ cal K^{-1} mol^{-1}. The enthalpy (H) and entropy (S) changes were obtained as -45.6 kJ mol^{-1} and -68.2 J K^{-1} mol^{-1}, respectively. The enthalpy change in the absorbing reaction of NO with an aqueous solution of Fe(II)–EDTA was reported to be -66.1 kJ mol^{-1} (-15.8 kcal mol^{-1}),[17] which is lower by 20.5 kJ mol^{-1} than that of CR–Fe(II). The strong NO-adsorbing capability of Fe(II)–EDTA is considered to be derived from the strong back donation from Fe(II) to the NO molecule due to the increase in the electron density of the iron atom. In the case of the CR–Fe(II) complex, however, the coordination of NO to the Fe(II) ion is relatively weak due to the electron-releasing effect of the benzyl group compared with the EDTA ligand.

6.3. Comparison of the Dry System with Aqueous Dispersion

The aqueous dispersion of CR–Fe(II) complex has been compared with the dry-type complex as an adsorbent of NO. Thus, both the adsorption rate and the complex formation constant K of the dry-type complex are larger than those of the aqueous dispersion. This high adsorbing capability observed for dry-type CR–Fe(II) complex is attributed to the direct coordination of NO to the active center of Fe(II) ions.

6.3.1. Adsorption rate The coordinating reaction of the aqueous dispersion is considered to occur by the following two steps[54]: (1) the dissolution of gaseous NO into the supernatant of the dispersion, and (2) the coordination of the dissolved NO to the Fe(II) in the resin. In the case of the dry-type CR–Fe(II) complex, the NO molecules in the gas phase directly coordinate to the Fe(II) ions in the complex. Thus, the adsorption of NO occurs more rapidly due to the shortcut of the first dissolution step, which is inevitable for the aqueous dispersion system. The fast adsorption by the dry-type complex is considered to be reasonable from the viewpoint of the difference between the NO concentration in the gas phase and the aqueous phase. The concentration of NO in the supernatant of the aqueous dispersion is calculated to be 1.93×10^{-6} mol dm^{-3} by means of Henry's law when the supernatant is equilibrated with 1000 ppm (4.08×10^{-5} mol dm^{-3}) of NO at 25°C. Thus the concentration of NO in the gas phase is about 20 times higher than that in the aqueous phase, which is favorable to the adsorption by the dry-type CR–Fe(II) complex.

6.3.2. Equilibrium constant The complex formation constant of the aqueous dispersion system according to Eq. (6) is calculated to be 31.1×10^3 dm^3 mol^{-1}.[54] On the other hand, the equilibrium constant K of the dry-type CR–Fe(II) system according to Eq. (2) is 205×10^3 dm^3 mol^{-1} when the dimension of the constant is adjusted to that of the dispersion system by considering the dilution of NO molecules in the gas phase. Thus, the equilibrium constant of the dry-type system is about 7 times larger than that of the aqueous dispersion system. In the case of the aqueous dispersion system, both water molecules and the IDA ligands coordinate on Fe(II) ions, and the coordination of the NO on the Fe(II) ions undergoes a ligand-exchange reaction by replacing the coordinated H_2O molecule with NO (cf. Fig. 5.28). In contrast, the Fe(II) ions in the dry-type CR–Fe(II) complex are thought to have some vacant coordination sites, probably because some of the coordinated H_2O molecules are removed by desiccation, and the ligand groups of the resin are not flexible enough to occupy the vacant sites. Thus the NO molecule will coordinate directly to the Fe(II) ion, which is considered to raise the equilibrium constant of the coordination of NO on the Fe(II) ion in the dry-type adsorbent.

Figure 5.28 Coordinatin of NO to the CR–Fe(II) complex in water.

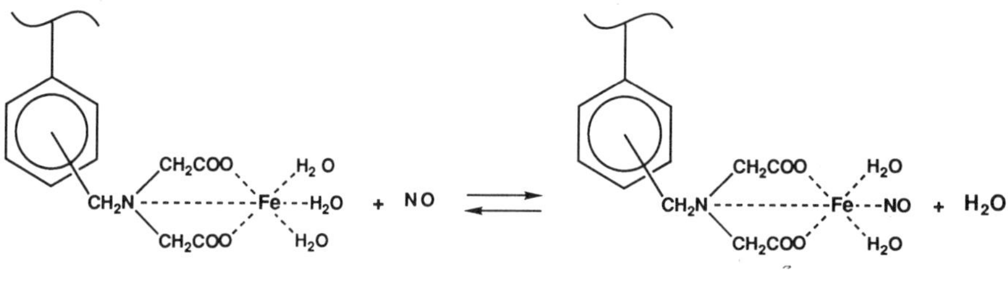

6.4. Recovery of the Resin

The solid part of the deactivated complex can be easily re-collected by centrifugation or filtration without a significant loss of the resin compared with the re-collection of IDA from an aqueous solution of the Fe(II)–IDA complex. The re-collected complex is regenerated by leaching the Fe ions with hydrochloric acid, followed by washing with an alkaline solution and then adding a fresh solution of $FeSO_4$. The adsorbent is completely regenerated without a significant decrease in the amount of adsorbed NO at equilibrium.[56]

6.5. Simultaneous Adsorption of Nitrogen Monoxide and Sulfur Dioxide

The simultaneous removal of NO and SO_2 is of practical importance because both gases are usually found together in flue gases. Figure 5.29 depicts adsorption curves of NO in the presence (closed circles) and absence (open circles) of SO_2. The presence of SO_2 does not suppress the adsorption of NO, but, on the contrary, a little promotes the smooth adsorption of NO.

Sulfur dioxide is removed by the aqueous adsorbent dispersions at the same time as NO. Up to 93% of the charged SO_2 can be absorbed at equilibrium within 35 min, as shown with open circles in Fig. 5.30. Thus the simultaneous removal of SO_2 and NO is successful using the present adsorbent dispersions. The observed increase in the adsorbed NO in the pres-

Figure 5.29 Adsorption of NO from 7 dm^3 of nitrogen gas containing 860 ppm of NO in the presence (●) and absence (○) of SO_2 by adsorbent dispersions prepared from 8.72 g of $FeSO_4$ and 7.53 g of the dried fine chelate resin in water.

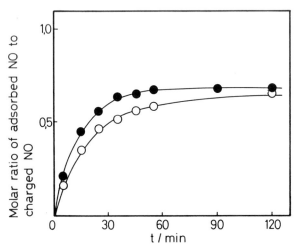

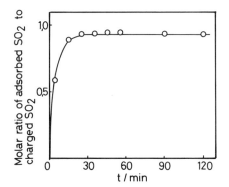

Figure 5.30 Adsorption of SO_2 from 7 dm³ of nitrogen gas containing 1180 ppm of SO_2 and 860 ppm of NO by the aqueous dispersion of CR–Fe(II) complex.

ence of SO_2 (Fig. 5.29) can be explained by the consideration that the dissolved SO_2 in an aqueous phase brings about HSO_3^- or SO_3^{2-} ions, which further react with NO in an aqueous phase,[45]

$$SO_3^{2-} + 2NO \rightleftharpoons SO_3(NO)_2^{2-} \rightarrow N_2O + SO_4^{2-}. \quad (9)$$

7. Other Inorganic Adsorbents for NO

Besides the polyamine N-carboxylato-Fe(II) complex system, the crystals of ferric hydroxide (such as jarosites) are also available as chemical adsorbents for NO[57–59]

7.1. Jarosites

Jarosites, stoichiometrically expressed as $KFe_3(SO_4)_2(OH)_6$, have a layer structure in the crystal, as shown in Fig. 5.31. This crystal can be prepared from $Fe_2(SO_4)_3$ and K_2SO_4 by mixing in hot water. Indeed the crystal shows an adsorption capability for NO, but its capability is rather low. The amount of adsorbed NO per g is 5.3×10^{-4} mmol at an NO concentration of 300 ppm at 100°C. This is because the surface area of the crystal is as low as 7.3 m²/g, even when it is crushed with a ball mil.[57] Thus in this case, an increase in the surface area is necessary for the efficient adsorption of NO.

7.2. FeOOH Systems

Iron(III) oxide hydroxides in α-, β-, or γ- form can also adsorb NO through chemical interaction of the NO molecule with O^{2-} on the crystal surface (Fig. 3.32). These crystals can easily be prepared from

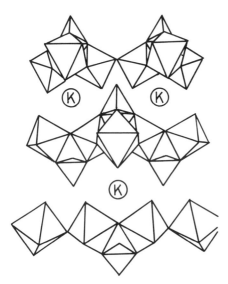

Figure 5.31 Layer structure of jarosite crystal. The centers of the octahedra are occupied by Fe and those of the tetrahedra by S.

$Fe_2(SO_4)_3$ and Na_2CO_3 by heat treatment. The α-FeOOH thus prepared acquires 60 m²/g of specific surface area, and adsorbs NO according to the Langmuir adsorption isotherm.[58] Moreover, α-FeOOH supported on an activated carbon fiber, whose surface area is 870 m²/g, acquires a twelvefold higher activity for NO at maximum compared with the pure α-FeOOH crystals. The activated carbon fiber supported the crystals can adsorb 0.043

Figure 5.32 Adsorption scheme of NO by FeOOH.

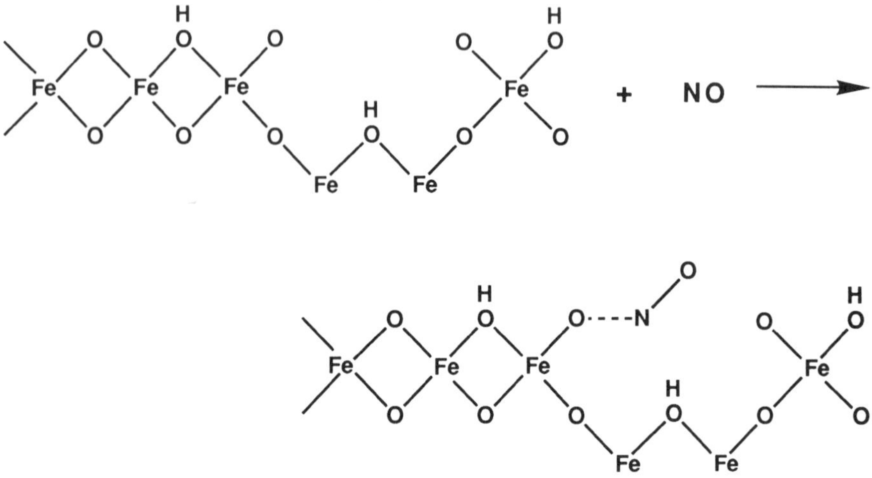

mmol of NO per g at 100°C at an equilibrated NO concentration of 300 ppm.[59] Thus the adsorbing capability is greatly controlled by the surface area of the adsorbents.

8. Concluding Remarks

1. The importance of porosity is demonstrated for polymer complexes used for gas separation. The examples are shown by a chelate-resin–Fe(II) complex for NO separation as well as a polystyrene-resin–$CuAlCl_4$ complex for CO separation, a hollow-fiber-supported Co(salen) complex for O_2 separation, and so on.
2. Chelate resin containing IDA moieties acquire high surface area by the following treatments: (a) immobilization of trivalent cations, (b) desiccation of the resin complex after washing with an organic solvent miscible with water. The increase in the porosity of the resin complex is attributed to the formation of a 1:1 complex formation of trivalent cation with IDA moieties.
3. The dry-type chelate-resin-immobilized Fe(II) complex can work as the adsorbent of NO through the complex formation of NO with Fe(II) ions. Its efficiency for the adsorption of NO is greatly dependent on surface area and is improved by such treatments as washing with methanol.
4. A mixed-valence complex of chelate resin with Fe(II) and Fe(III) ions has a high efficiency for the adsorption of NO compared with the CR–Fe(II) complex, due to the high surface area caused by the incorporated Fe(III) ions.
5. The CR–Fe(II) complex can also work as a wet-type adsorbent for NO and can simultaneously remove both sulfur dioxide and NO.

REFERENCES

1. A. L. Kohl and F. C. Riesenfeld, *Gas Purification*, 3rd ed., Gulf Publishing Co., Houston (1979).
2. D. M. Ruthven, *Principles of Adsorption and Adsorption Process*, Wiley, New York (1984).
3. D. J. Haase and D. G. Walker, *Chem. Eng. Progr.* **70**, 74 (1974).
4. D. G. Walker, *Chemtech* **5**, 308 (1975).
5. D. J. Haase, *Chem. Eng.* **82**(16), 52 (1975).
6. H. Hirai, S. Hara, and M. Komiyama, *Bull. Chem. Soc. Jpn.* **59**, 1051 (1986).
7. H. Hirai, S. Hara, and M. Komiyama, *Bull. Chem. Soc. Jpn.* **60**, 385 (1987).
8. A. T. Larson and C. S. Teitsworth, *J. Am. Chem. Soc.* **44**, 2878 (1922).
9. H. Hirai, K. Wada, K. Kurima, and M. Komiyama, *Angew. Makromol. Chem.* **142**, 105 (1986).

10. H. Hirai, K. Wada, K. Kurima, and M. Komiyama, *Bull. Chem. Soc. Jpn.* **59**, 2553 (1986).
11. H. Hirai, K. Kurima, and M. Komiyama, *Chem. Lett.* **1986**, 671 (1986).
12. H. Hirai, K. Kurima, K. Wada, and M. Komiyama, *Chem. Lett.* **1985**, 1513 (1985).
13. H. Hirai, S. Hara, and M. Komiyama, *Chem. Lett.* **1986**, 257 (1986).
14. H. Hirai, S. Hara, and M. Komiyama, *Angew. Makromol. Chem.* **130**, 207 (1985).
15. R. H. Bailes and M. Calvin, *J. Am. Chem. Soc.* **69**, 1886 (1947).
16. E. Tsuchida, H. Nishide, M. Ohyanagi, and H. Kawakami, *Macromolecules* **20**, 1907 (1987).
17. Y. Hishinuma, R. Kaji, H. Akimoto, F. Nakajima, T. Mori, T. Kamo, Y. Arikawa, and S. Nozawa, *Bull. Chem. Soc. Jpn.* **52**, 2863 (1979).
18. K. Ogura and T. Ozeki, *Electrochim. Acta* **26**, 877 (1981).
19. N. Toshima, H. Asanuma, and H. Hirai, *Chem. Lett.* **1986**, 667 (1986).
20. N. Toshima, H. Asanuma, and H. Hirai, *Bull. Chem. Soc. Jpn.* **62**, 893 (1989).
21. H. Asanuma and N. Toshima, *J. Polym. Sci. Polym. Chem. Ed.* **28**, 907 (1990).
22. D. Sellmann, *Angew. Chem., Int. Ed. Engl.* **10**, 919 (1971).
23. Y. Kurimura, F. Ohta, J. Gohda, N. Nishide, and E. Tsuchida, *Macromol. Chem.* **183**(12), 2889 (1982).
24. R. Kunin, *J. Polymer Sci. C* **16**, 1457 (1967).
25. R. Kunin, *J. Polymer Sci. B* **2**, 587 (1962).
26. E. Borlel, *Przemysl Chem.* **44**, 255 (1965).
27. R. Kunin, *J. Polymer Sci. A-1* **6**, 2689 (1968).
28. I. M. Abrams, *Ind. Eng. Chem.* **48**, 1469 (1956).
29. J. Seidl and J. Malinsky, *Chem. Prumysl* **13**, 100 (1963).
30. H. Nishide and E. Tsuchida, unpublished results.
31. K. Häupke and V. Pientka, *J. Chromatogr.* **102**, 117 (1974).
32. P. P. Wieczorek, B. N. Kolarz, and H. Galina, *Angew. Makromol. Chem.* **126**, 39 (1984).
33. B. N. Kolarz, P. P. Wieczorek, and M. Wojaczyńska, *Angew. Makromol. Chem.* **96**, 193 (1981).
34. J. Hradil and F. Švec, *Angew. Makromol. Chem.* **130,** 81 (1985).
35. B. N. Kolarz and P. P. Wieczorek, *Angew. Makromol. Chem.* **96**, 201 (1981).
36. H. Asanuma and N. Toshima, *J. Chem. Soc., Chem. Commun.* **1989**, 1075 (1989).
37. H. Burrell, *Polymer Handbook*, J. Brandrup and E. H. Immergut, Eds., Wiley-Intersci. Pub., New York (1974), Chap. 4, p. 337.
38. D. O'Sullivan, *Chem. Eng. News* **66**(40), 8 (1988).
39. I. Mochida, K. Suetsugu, H. Fujitsu, and K. Takeshita, *J. Catal.* **77**, 519 (1982).
40. T. Shikada, T. Oba, K. Fujimoto, and H. Tominaga, *Ind. Eng. Chem., Prod. Res. Rev.* **23**, 417 (1984).
41. G. Tuenter et al.., *Ind. Eng. Chem. Prod. Res. Dev.* **25**, 633 (1986).
42. H. Juentgen, E. Richter, and H. Kuehl, *Fuel* **67**, 775 (1988).

43. H. Juentgen, E. Richter, K. Knoblauch, and T. Hoang-Phu, *Chem. Eng. Sci.* **43**, 419 (1988).
44. P. Davini, *Fuel* **67**, 24 (1988).
45. S.-G. Chang, D. Littlejohn, and S. Lynn, *Environ. Sci. Technol.* **17**, 649 (1983).
46. E. Sada, H. Kumazawa, and H. Machida, *Ind. Eng. Chem. Res.* **26**, 2016 (1987).
47. E. Sada, H. Kumazawa, and Y. Yoshikawa, *AIChE J.* **34**, 1215 (1988).
48. C. C. McDonald, W. D. Phillips, and H. F. Mower, *J. Am. Chem. Soc.* **87**, 3319 (1965).
49. D. M. Ruthven, *Principles of Adsorption and Adsorption Process*, Wiley, New York (1984), p. 49.
50. E. Sada, H. Kumazawa, and H. Machida, *Ind. Eng. Chem. Res.* **26**, 1468 (1987).
51. T. Sato, T. Goto, T. Okabe, and F. Lawson, *Bull. Chem. Soc. Jpn.* **57**, 2082 (1984).
52. H. Hirai, H. Asanuma, and N. Toshima, *Chem. Lett.* **1985**, 655 (1985).
53. H. Hirai, H. Asanuma, and N. Toshima, *Chem. Lett.* **1985**, 1921 (1985).
54. N. Toshima, H. Asanuma, K. Yamaguchi, and H. Hirai, *Bull. Chem. Soc. Jpn.* **62**, 563 (1989).
55. K. Pearsall and F. T. Bonner, *Inorg. Chem.* **21**, 1978 (1982).
56. H. Asanuma, A. Takemura, N. Toshima, and H. Hirai, *Ind. Eng. Chem. Res.* **29**, 2267 (1990).
57. K. Inouye, I. Nagumo, K. Kaneko, and T. Ishikawa, *Z. Phys. Chem., Neue Folge* **131**, 199 (1982).
58. T. Hattori, K. Kaneko, T. Ishikawa, and K. Inouye, *Nippon Kagaku Kaishi* **1979**, 423 (1979).
59. K. Kaneko, S. Ozeki, and K. Inouye, *Nippon Kagaku Kaishi* **1985**, 2315 (1985).

6
Polymer Complex Membranes for Gas Separation

Hiroyuki Nishide and Eishun Tsuchida

1. Introduction
2. Liquid Membranes for Facilitated Transport
3. Selective Sorption in Polymer Complex Membranes
4. Solid Membranes for Facilitated Transport
5. Transport Mechanism and the Effects of Fixed Complexes
6. Conclusions and Further Studies

1. Introduction

A polymer–metal complex is composed of a synthetic polymer and transition-metal ions. Its synthesis represents an attempt to give an organic polymer inorganic functions.[1] One of the characteristic chemical functions of metal ions and their complexes is the specific and reversible binding of gaseous molecules. Typical examples are the efficient oxygen carriage and storage by hemoglobin and myoglobin, respectively, in a living body. Hemoglobin and myoglobin are conjugated proteins with (an) iron porphyrin(s), of which central iron(II) coordinately and selectively binds molecular oxygen, as shown in Eq. (1).

On exposure of the hemoglobin or myoglobin solution to oxygen, the iron porphyrin complex (deoxy form, dark red) transfers to its oxygen-coordinated complex ($[O_2]:[Fe] = 1:1$, oxy form) and turns brilliant red.

$$\text{[Fe porphyrin with imidazole, 2 C}_2\text{H}_4\text{COOH]} + O_2 \rightleftharpoons \text{[Fe porphyrin}\cdot O_2 \text{ with imidazole, 2 C}_2\text{H}_4\text{COOH]} \tag{1}$$

This oxygen-binding reaction is rapid and reversible in response to partial oxygen pressure.

For oxygen to be reversibly bound, the oxidation potential of a complex such as iron porphyrin must be in a range such that a certain amount of electron charge is donated to the coordinated oxygen molecule, yet not so great that irreversible oxidation of the metal occurs.[2] (This feature is also adequate for the reversible binding of other gaseous molecules to metal ions.) A subtle change in the ligands attached to the central metal results in a modification of the metal oxidation potential and therefore in the stability of the oxygen-coordinated complex. In the case of iron porphyrin as an oxygen carrier, the iron must be bonded to five electron-donating atoms to form a five-coordinated deoxy structure before it can reversibly bind molecular oxygen. The protein chains of hemoglobin and myoglobin have globular and compact conformations, and the iron porphyrin is embedded and immobilized in the pocketlike inside space of the globular protein. The iron porphyrin bonds to the imidazolyl group of the histidine residue of the pocket-forming segment of the protein to satisfy the five-coordinated structure and becomes the oxygen-binding site [Eq. (1)]. The hydrophobic character of the space inside the pocket, caused by the surrounding hydrophobic amino acid residues, also enhances the oxygen-binding function of hemoglobin and myoglobin.

The oxygen-binding affinity or oxygen-binding equilibrium constant (K) of the reaction [Eq. (1)] is extremely large for myoglobin ($1/K = 0.5$ mm Hg) to store oxygen and moderately small ($1/K = 27$ mm Hg) for hemoglobin to deliver oxygen at terminal tissues. 100 ml of human blood contains 16 g hemoglobin and absorbs 23 ml of oxygen when exposed to air. This absorption amount is ~60 times the volume (0.4 ml) physically dissolved in water. This extraordinarily high capacity and efficiency in oxygen transporting by blood are attributed to the specific, rapid, and reversible oxygen binding to the iron porphyrin complex coexisting with the macromolecular protein.

Metal complexes with selective and efficient oxygen-carrying capability (oxygen carriers) like hemoglobin are expected to yield new systems for oxygen separation and transportation. However, for example, the iron porphyrin complex isolated from hemoglobin is immediately and irreversibly oxidized to its ferric [iron (III)] form, and it does not act as a reversible oxygen-binding site. The protein of hemoglobin occludes the iron porphyrin complex not only to satisfy the five-coordinated deoxy form but also to provide the molecular environment to maintain the oxygen-binding capability of the complex under a wide range of conditions. The metal complexes bound to synthetic polymers often show specific types of chemical reactions, because the reactions are affected by the polymers that surround the metal complex moieties.[3] Indeed it has been shown that synthetic polymer–metalloporphyrin complexes can specifically, rapidly, and reversibly bind molecular oxygen from air.[2,3]

In this chapter, specific and reversible bindings of small molecules from gaseous mixtures to the polymer–metal complexes in the membrane states are described, by especially discussing selective permeation or transport of the gaseous molecules across the membranes of polymer–metal complexes. Physicochemical aspects and application potentials of the polymer-complex membranes for gas separation are also described.

2. Liquid Membranes for Facilitated Transport

Hemoglobin in blood dynamically circulates through the whole body and delivers oxygen from the lungs to terminal tissues. On the other hand, the hemoglobin solution had first been examined for gas separation by the procedure in which the hemoglobin solution is retained within the pores of a microporous membrane.[4] The process for oxygen transport with hemoglobin across a liquid membrane can be explained by using scheme (a) in Fig. 6.1.

A mobile carrier of oxygen, here hemoglobin, is dissolved in a liquid membrane that separates the upstream (feed-air stream) from the downstream (product stream). The upstream is maintained at a sufficiently high oxygen pressure that the oxygen carrier is in its oxy form at the upstream–membrane interface. The downstream is kept at a sufficiently low oxygen pressure (by being evacuated or with sweep inert gas) to maintain the carrier in its deoxy form at the membrane–downstream interface. The oxygen carrier thus acts as a shuttle, picking up oxygen at the upstream–membrane interface, diffusing across the membrane as the oxy form, releasing oxygen to the downstream and then diffusing back to the upstream–membrane interface to repeat the process. Because the carrier is specific for oxygen, the rate of oxygen transport is enhanced with no effect on the rate of nitrogen transport, resulting in a considerably higher oxygen enrichment of the downstream than is possible in the absence of the carrier. Such a process of selective transport across membranes facilitated by carriers is called *facilitated transport* or *carrier-mediated transport*.[5]

A Millipore membrane holding a hemoglobin solution was able to dem-

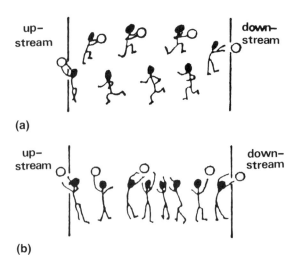

Figure 6.1 Scheme for facilitated transport of gaseous molecule with a carrier (complex) through a membrane. (a) Liquid membrane with a mobile carrier; (b) solid membrane with a fixed carrier.

onstrate an eightfold increase in oxygen flux over that attainable through a pure water layer of the same thickness, due to the specific oxygen binding to hemoglobin.[4,6] This represented an oxygen/nitrogen selectivity of ~14. A similar facilitation effect through the porous membrane containing a hemoglobin solution was reported for the transport of carbon monoxide.[7] However, hemoglobin is highly unstable in aqueous solutions outside the body, and is too bulky (mol. wt. ~64,500) to diffuse rapidly across the membrane.

The first example of facilitated transport liquid membranes consisting of a synthetic metal complex as a mobile oxygen carrier within a microporous membrane was made by using bis(histidine)cobalt (structure *1*).[8] The best result was an approximate doubling of the oxygen flux over that through water and a resulting selectivity of 3.5.

1

[Structure 2: Cobalt Schiff-base complex with salicylidene groups, =N—Co—N= core, bridging (CH₂)₃—N(H)—(CH₂)₃]

2

The class of facilitated-transport liquid membranes for the production of oxygen-enriched air has been much improved and established by the group at Bend Research Inc.[9] They used a series of cobalt Schiff-base complexes as the mobile oxygen carriers in liquid membranes,[10] and three important types of complexes are as follows: [N,N'-bis(salicylidene)-n-propyldipropylenetriamine]cobalt (structure 2), [N,N'-bis(3-methoxysalicylidene)tetramethylethylenediamine)] cobalt (structure 3), and the cobalt dry-cave complex synthesized by Busch[11] (structure 4). In 2 the five-coordinated deoxy form is supplied by the cyclic ligand itself. 3 and 4, on the other hand, have only four coordinating atoms in their ligand, and the fifth coordinating atom is to be supplied by an axial base externally added to the complex solution, such as 1-methylimidazole and pyridine. Liquid membranes were prepared by immersing a microporous membrane into a solution of the complex. The solution is held strongly by capillary action within the pores. A typical example of microporous membranes is Ultipor NM nylon 6,6 with 125 μm thickness, 0.2 μm pore diameter, and 0.80 porosity. Porous cellulose acetate and Gortex polytetrafluoroethylene membranes are also available. The Ultipor membrane containing a γ-butyrolactone solution of structure 3 with 4-dimethylaminopyridine at $-10°C$ exhibited oxygen/nitrogen selectivity in excess of 20, and a permeate stream with oxygen purity in the range of 80–90% was achieved using 1-atm pressure upstream and a low-pressure downstream. An oxygen permeability coefficient $P_{O_2} = 2.6 \times 10^{-8}$ cm³ (STP) cm cm^{-2} sec^{-1} (cm Hg^{-1}) [(unit permeate gas volume)(unit membrane thickness)(unit membrane area)$^{-1}$ (unit time)$^{-1}$(unit pressure difference)$^{-1}$] about one-half that of a silicone rubber membrane was observed.[10,13]

[Structure 3: Cobalt Schiff-base complex with 3-methoxysalicylidene groups (OCH₃, H₃CO), =N—Co—N= core, bridging H₃C—C(CH₃)—C(CH₃)—CH₃]

3

4

The oxygen flux and permselectivity are related to critical fundamental parameters of the membrane system, that is, concentration of the carrier solubilized in the liquid medium, diffusivity of the carrier, upstream and downstream pressure, oxygen-binding affinity and kinetics of the carrier, operation temperature, and membrane thickness.[9,10] First the carrier should be present at a high concentration in the liquid medium; both clearly need to be relatively involatile. The effect of the liquid medium was also described by an inverse correlation result that the oxygen permeability of the liquid membranes containing structure 3 decreased in the order dimethylformamide > benzonitrile > dimethylacetamide > butyrolactone > benzaldehyde > 1-methylpyrrolidone > propylene carbonate with an increase in solvent viscosity, which corresponds to an inverse correlation between the viscosity of the liquid medium and the diffusivity of the carrier. One of the best liquid media was butyrolactone, which has a vapor pressure at 25°C of only ~0.1 cm Hg, and its viscosity is ~2 cP. The facilitated transport increased with the complex concentration in the liquid medium, for example, up to $0.4M$ carrier in butyrolactone, and the oxygen/nitrogen selectivity increased from 26 to 45, simply because there was more carrier available for oxygen transport. But above the optimum concentration, the solution viscosity of the liquid medium increased noticeably, and the diffusivity of the relatively large oxygen complex fell. The carrier must be small and relatively mobile in the liquid medium (Fig. 6.2).

The pressure at which the downstream side of a membrane process is operated is important to the process economics, because the cost of compression increases with decreasing downstream (product) side pressure

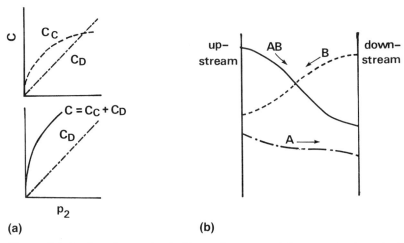

Figure 6.2 Dual-mode sorption in (a) membrane containing a carrier (complex) and (b) a concentration profile for the species in a liquid membrane. C: concentration of sorbed gas; C_D: concentration of sorbed gas according to Henry's law; C_C: concentration of sorbed gas according to Langmuir's isotherm; p_2: atmospheric or upstream partial pressure of gas; A: dissolved gas; B: unbound carrier (complex); AB: carrier–gas complex.

even though product purification increases. The change in P_{O_2} derives from the nature of the oxygen sorption isotherm of the carrier, which is represented by Langmuir's isotherm (Fig. 6.2). The oxygen-binding affinity [K in Eq. (2)] should be chosen to yield the maximum degree of oxygen binding on the upstream side, with a concomitant minimum degree of oxygen binding on the downstream side. There is an optimum oxygen-binding affinity (K_{opt}) at which there is a maximum difference in the degree of oxygen binding.

$$\text{SCo} + \text{O}_2 \rightleftharpoons \text{SCo–O}_2 \qquad (2)$$

$$K_{opt}^{-1} = \sqrt{p_1, p_2}, \qquad (3)$$

where CoS represents a cobalt Schiff-base complex, and p_1 and p_2 refer to the pressures on the two sides of the membrane. For typical conditions, $p_1 = 15$ cm Hg (oxygen in 1 atm air) versus $p_2 = 1$ cm Hg, the optimum affinity occurs at $K = 0.25$ cm Hg^{-1} with 60% of the transport difference (80% oxygen binding at the upstream and 20% at the downstream). Carriers with such an affinity are the most desirable for use in facilitated transport.

In polymeric membranes, gas permeability normally increases with temperature, and the selectivity is usually almost independent of temperature. In facilitated-transport membranes, however, the facilitation efficiency is affected by temperature dependencies of the binding affinity and the kinetics of the carrier and goes through a maximum with temperature, because the former is enlarged at lower temperature and the latter increases with temperature.

The temperature effect of the liquid membranes for facilitated transport should be also discussed from the viewpoint that the most difficult requirement of such membranes is of long-term stability.[12] Three factors contribute to the decline in membrane performance: (1) irreversible oxidation of the carrier through the repeated oxy–deoxy cycle, (2) evaporative loss of the liquid medium and sometimes of the externally added axial base ligand, and (3) poisoning of the carrier by minor constituents of the normal atmosphere, that is, water vapor and carbon dioxide. To overcome these factors membrane operation at low temperature is recommended: for example, the 3/butyrolactone membrane maintained its oxygen product purity at ~70% over the first several weeks at $-10°C$ operation.[10]

Many attempts have been reported of oxygen transport through liquid membranes by modifying the above-mentioned factors to advance the facilitation efficiency. Photografted cellulose membrane was applied as the supporting porous membrane of a DMF solution of N,N'-disalicylidene-ethylenediaminecobalt (structure 5, CoS), which gave relatively large fluxes with a very advantageous oxygen/nitrogen selectivity of ~50 in spite of its short lifetime.[13] The cobalt Schiff-base complex structure 4 has a cyclidene lacunar structure, presented in structure 6, of which the distorted coordination sphere improves its kinetic parameters in oxygen binding. The dicyanobenzene solution of structure 4 supported with Millipore porous polyvinylidenefluoride membrane yielded 78% of oxygen-enriched air at $0°C$.[14]

To reduce the molecular size of a carrier and increase its diffusivity, a series of cobalt complexes was tested as oxygen carriers of liquid membranes for air separation. For example, tripropyrene tetramineimidazolethiocyanito cobalt in DMSO membrane showed an oxygen/nitrogen selectivity of 33.[15] The regeneration properties of deteriorated carriers were also examined.[16] The liquid medium, containing degraded-carrier structure 5, worked for several days and was treated with reductants such as zinc at an elevated temperature, which regenerated the permselectivity of the membrane up to that of the virgin carrier.

For carbon monoxide transport, the following iron (II) complex was reported as the mobile carrier in liquid membranes[17]

$$(L)(C_6H_5CN)_2Fe + CO \rightleftharpoons (L)(C_6H_5CN)Fe-CO + C_6H_5CN. \quad (4)$$

Here (L) is a macrocyclic Schiff base. The transport mechanism was also discussed. Cuprous chloride solutions are known to uptake carbon monoxide efficiently (Chap. 7). Facilitated carbon monoxide transport was also reported for the DMSO liquid membrane containing both copper–imidazole complex and benzoin, in which the Cu(I) complex formed through the reduction of copper with benzoin selectively binds carbon monoxide, to show a CO/nitrogen selectivity of ~28.[18]

The research group at Air Products recently reported a new class of facilitated gas-separation membranes consisting of a gas-reactive molten salt held within a porous support.[19,20] This concept has been used to fabricate membranes that, at high temperatures ~300°C, selectively permeate ammonia from mixtures of ammonia, nitrogen, and hydrogen. The active membrane layer is a film of molten zinc chloride that undergoes the following selective and reversible binding reaction:

$$Zn^{2+} + n\, NH_3 \rightleftharpoons Zn(NH_3)_n^{2+}, \quad (5)$$

$n = 1$ or 2. The zinc ion acts as a facilitated-transport carrier for ammonia, while the melt is essentially a barrier to noninteractive gases. This results in rapid permeation of ammonia [$P = 10^{-5}$ cm³ (STP) cm cm^{-2} s^{-1} (cm Hg^{-1}) at 300°C] with high ammonia selectivity (1000–3000). High-temperature molten salt membranes have several unique features, including a high carrier concentration, favorable diffusivities and reaction rates, and low volatility, and a study to prepare their membranes for practical tests has been started.

It is known that aqueous solution of silver nitrate absorbs olefines selectively. The silver nitrate solution retained in a hollow fiber module has been examined for production of 98% propylene.[21] The Teflon porous-membrane-supported system gave an ethylene/ethane selectivity of ~700, although the performance declined with time due to water vaporization and deposition of silver.[22] Nafion perfluorosulfonic acid membrane, of which cation was exchanged with silver ion, showed facilitated transport of ethylene with $P = 10^{-7}$ cm³ (STP) cm cm^{-2} s^{-1} (cm Hg^{-1}).[23] In this membrane the carrier silver ion is immobilized onto the sulfonate anion residue of the polymer matrix and acts as an immobilized carrier while the membrane retains water in the micropore.

As mentioned above, liquid membranes holding metal complexes for facilitated transport could provide an elegant method for gas separation with the following advantages:

1. Improved selectivity;
2. Increased flux; and
3. Expensive carriers can be used.

But there still remain the following unresolved issues:

1. Evaporative loss of the liquid medium;
2. Chemical instability of the carrier;
3. Temperature limitation for use;
4. Membrane thickness;
5. Solubility limitation of the carrier in the liquid medium; and
6. Lower mobility of the carrier than that of permeate gaseous molecules.

To overcome these problems with liquid membranes, solid membranes in which the carrier is fixed in the solid state are expected to be promising candidates.

3. Selective Sorption in Polymer Complex Membranes

Except for a few studies, only a relatively modest effort has been directed towards solid (absolutely solvent-free) membranes containing fixed carriers for facilitated transport. The reasons are probably as follows:

1. Inactivation of the carrier after fixation in the solid state;
2. Nonuniformity in chemical reactivity of the fixed carrier;
3. Defect formation in the solid membrane; and
4. Low diffusivity of a permeate molecule via the fixed carrier.

The first serious issue was how to maintain the activity of the carrier even after the fixation in the solid-state membrane.

The complex often loses its activity to bind a gaseous molecule rapidly after fixation in a dry polymer membrane. A solvent or a polymer residue occupies the sixth coordination site (binding site) of the complex (carrier) during the membrane preparation (casting) procedure and/or after the immobilization. A permeate molecule has to displace the occupant, and the reaction becomes very slow. If the gaseous-molecule-binding site of the carrier complex is kept vacant after the dry membrane preparation, this carrier reacts very rapidly even in the solid state. By a combination of chemical modification of the complex and polymer attachment of the complex, we recently succeeded in preparing solid membranes of polymer–metalloporphyrin complexes, whose sixth coordination site is vacant even in the solid state and able to bind molecular oxygen very rapidly.[24–29]

A series of cobalt– and iron–porphyrin complexes, tetrakis(alkylamidophenyl)porphinatometals (structures *7–11*), was synthesized as the oxygen-binding site to be fixed in the polymer membranes.[24–26] For example, a solution of complex *7* was carefully cast under an absolutely oxygen-free atmosphere, followed by drying in vacuo, to yield a flexible, transparent, and wine-red-colored membrane. These modified cobaltporphyrin complexes were successfully fixed in polymer matrices with their oxygen-binding capability by coordinate bonding with poly(1-vinylimidazole- or 4-vinylpyridine-co-alkylmethacrylate) as an axial base ligand (structures *7–10*) or by covalent bonding (structure *11*), because the complexes have a steric pocketlike cavity on the porphyrin plane to keep their oxygen-coordinating site vacant before oxygen binding even in the solid or solvent-free state. This stereostructure of the porphyrin complex is one of the key points in preparing an active carrier complex in the solid state.

Selective and reversible oxygen binding to the fixed complexes was first confirmed with microgravimetric measurements (Fig. 6.3).[27] For example, the structure *7* membrane containing 30% cobaltporphyrin (CoP) complex moiety sorbed ~4 ml oxygen/g polymer, which is more than 500 times larger than that of physically dissolved nitrogen. This extraordinarily large amount of dissolved oxygen in the polymer membrane is based on the specific and reversible oxygen binding to the CoP complex in the polymer. Figure 6.3 also indicates that the oxygen sorption is in response to an at-

R: — CH_3

— C_8H_{17}

— $C_{12}H_{25}$

7

8

9

10

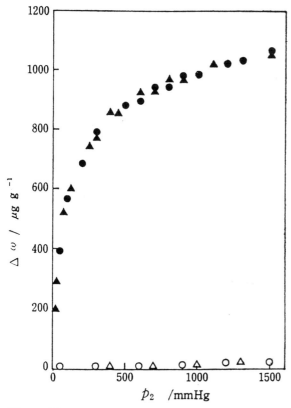

11

Figure 6.3 Oxygen sorption into the membrane of the polymer cobaltporphyrin complex (structure 7). Δω : weight increase of the membrane or gas-sorption amount measured with a microbalance (sorbed gas weight per membrane unit weight); p_2: atmospheric pressure; closed symbols: oxygen sorption; open symbols: nitrogen sorption; Δ : repeated data.

mospheric oxygen pressure and is according to Henry's law for the physical sorption and Langmuir isotherm for the chemical dissolution to the complex, to give a physical solubility coefficient k_D of the gaseous molecule and an equilibrium constant K for the binding reaction [Eq. (6)], respectively.

$$\text{ImPCo} + O_2 \underset{k_{\text{off}}}{\overset{k_{\text{on}}}{\rightleftharpoons}} \text{ImPCo-}O_2, \quad K = k_{\text{on}}/k_{\text{off}}. \tag{6}$$

Here, PCo and Im represent cobaltporphyrin and imidazolyl residue, respectively. Examples of the K values are given in Table 6.1, which indicates that the oxygen-binding affinity of the CoP complex can be controlled with the chemical structure of the complex.[26,29]

Selective oxygen binding to the CoP complex moiety was confirmed in the in situ membrane state with general spectroscopy, because the membranes are transparent and flexible. For example, by sticking a membrane fragment to the cell window of a simple ir spectrometer, strong ir absorption at 1150 cm^{-1} for $^{16}O_2$ and 1060 cm^{-1} for $^{18}O_2$ attributed to an endon-type coordination of dioxygen to the metal ion appeared with an increase of partial oxygen pressure.[25]

The color of the membrane changes from dark red for veinous blood to brilliant red for arterial blood on exposure of the CoP membrane to oxygen atmosphere. This color change of the membrane also becomes a good probe of oxygen binding and can be monitored easily with visible absorption spectrometry. The oxy–deoxy spectral change was reversible in response to partial oxygen pressure with isosbestic points.[28] This is crucial evidence that the cobalt complex acts as an effective oxygen-binding site from the equilibrium viewpoint, even after fixing in the solid membrane.

It is known that the metalloporphyrin-coordinated oxygen is photodissociated under flash irradiation [Eq. (7)] and that the rapid oxygen-binding reaction can be analyzed. Photodissociation and recombination of the coordinated oxygen from and to the CoP complex in the solid membrane was observed by improving pulse and laser flash spectroscopic techniques.[24]

Table 6.1 Rate and Equilibrium Constants and Thermodynamic Parameters for the Oxygen Binding to the Cobaltporphyrin Compex in Solid-State Membrane at 25°C

Complex	$10^{-3}K$ (M^{-1})	ΔH (kcal mol^{-1})	ΔS (e.u.)	$10^{-7}k_{\text{on}}$ (M^{-1} s^{-1})	$10^{-3}k_{\text{off}}$ (s^{-1})
7	3.0	−13	−37	0.98	3.2
8	4.3	−13	−37	1.4	3.2
9	4.0	−13	−38	2.3	5.7
10	0.51	−11	−36	3.4	66
7[a]	0.92	−13	−40	0.0063	0.068

[a]Toluene solution.

The laser flash was applied perpendicular to the light path of the spectrophotometer, and the membrane was placed at the crossing of the laser flash and the light path at 45° to both. An example of the recombination time curve of oxygen is shown in Fig. 6.4(a), which shows that the reaction is completed within microseconds and is surprisingly rapid

$$\text{ImPCo-O}_2 \xrightarrow{h\nu} \text{ImPCo} + \text{O}_2 \xrightarrow{k_{app}} \text{ImPCo-O}_2. \qquad (7)$$

The surrounding oxygen concentration of the complex moiety in the membrane was estimated by multiplying atmospheric oxygen pressure around the membrane by physical solubility coefficient k_D of oxygen. The

Figure 6.4 Recombination reaction of the photodissociated oxygen to the complex fixed in the membrane. (a) Spectral change after laser flash irradiation; (b) plots of apparent oxygen-binding rate constant (k_{app}) versus oxygen concentration around the complex moiety in the membrane (p'); (c) scheme for recombination of the dissociated oxygen.

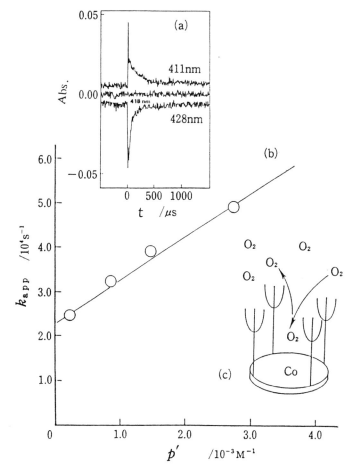

plots of the apparent oxygen-binding rate constant k_{app} in Eq. (7) versus the oxygen concentration is shown in Fig. 4(b): The linear relationship indicates a second-order kinetics for the oxygen-binding to the complex fixed in the dry polymer membrane similar to that in a hemogeneous solution system. That is, it was denied that the photodissociated oxygen directly recombines to the original CoP complex [Fig. 6.4(c)].

k_{on} and k_{off} [Eq. (6)] were listed in Table 6.1.[29] The CoP complex is kinetically active for oxygen-binding even in the solid membrane, if one carefully pays attention on the combination of complex structure and polymer species during introduction of the complex. In Table 6.1 one notices that the k_{on} and k_{off} values are rather larger than those of the corresponding complex in homogeneous toluene solution. In solution the binding site of the carrier complex is weakly but certainly solvated with the solvent molecule, so that the binding molecule has to kick out the occupant solvent.[30] On the other hand, the CoP complexes with cavitylike stereostructures, such as complexes 7 and 11, maintain their sixth coordination site vacant, and the gaseous molecule-binding reaction becomes its idealized one and is very rapid.

Table 6.1 also shows that the porphyrin structure clearly influences the oxygen-binding character of CoP and that the reduction in the steric hindrance on the porphyrin plane or the oxygen-binding and -dissociation pathway enhances the rate constants k_{on} and k_{off}. From the kinetic viewpoint in the oxygen-binding reaction, complex 10 is of great promise for facilitated transport through the membrane.

The deoxy–oxy cycle was repeated at 1-min intervals and recorded 1.2×10^5 times (~6 months) by changing the atmosphere of the membrane from reduced pressure (~2 cm Hg) to laboratory air pressure (76 cm Hg) at 25°C.[31] Lifetime or half-life periods for the oxygen-binding ability are summarized in Table 6.2 with the reference data of hemoglobin and a solution state of the corresponding complex. Hemoglobin in red blood works as an oxygen carrier for 4 months, because the cell encapsulates hemoglobin with a small amount of met-hemoglobin reductase or a reductant to repair irreversibly oxidized hemoglobin.[30] The stripped and isolated hemoglobin in aqueous solution works as an oxygen carrier only for 3 days.[30] Synthetic CoP complex 7 in hemogeneous toluene solution is able to repeat the oxy–deoxy cycle with a lifetime of ~1 day.

Table 6.2 Lifetime of the Metalloporphyrin Complexes as on Oxygen Carrier at Room Temperature

Complex	Physical State	Lifetime
hemoglobin	red cell	4 mo
hemoglobin	aqueous soln.	3 d
7	toluene soln.	1 d
7	solid membr.	> 6 mo

It is known that a metalloporphyrin complex loses its oxygen-binding ability through the following irreversible oxidation reactions, which are explained using a CoP complex[2,32,33]:

1. The oxy-CoP forms with another deoxy-CoP, a μ-dioxo CoP, which is converted irreversibly to Co(III)P without oxygen-binding ability, as represented here:

$$\tag{8}$$

$$\text{Co}^{II}(O_2)(B) + \text{Co}^{II}(B) \rightleftharpoons \text{Co}^{II}(B)\text{-O-O-Co}^{II}(B) \rightarrow \text{Co}^{III}(B)\text{-O}^{2-}\text{-Co}^{III}(B)$$

2. Water molecule (proton) attacks the CoP-bound dioxygen, to yield hydrodioxy radical and Co(III)P:

$$\tag{9}$$

$$\text{Co}^{II}(O_2)(B) + H_2O \rightarrow \text{Co}^{III}(OH^-)(B) + HO_2^{\cdot}$$

3. Although the CoP complex with an axial base ligand (B) is thermodynamically stable with a very large formation constant, its association and dissociation reactions are very rapid, and the CoP complex is labile in substitution reactions even with the axial base ligand. A moment dissociation of the axial base ligand in the oxygen-coordinated CoP complex causes electron transfer from the cobalt to the coordinated oxygen also to yield CO(III)P

$$\tag{10}$$

$$\text{Co}^{II}(O_2)(B) \rightarrow \text{Co}(O_2) + B \rightarrow \text{Co}^{III}(B) + O_2^{-}$$

For the CoP bound to a polymer ligand in the solid state, the irreversible oxidation via Eq. (8) is inhibited because CoP is molecularly dispersed and immobilized onto the polymer chain. The hydrophobic environment

12

[Structure 12: copolymer with CH₃, C(=O)OC₈H₁₇ side group and cyclopentadienyl-Mn(CO)₂(N₂) with CH₃ substituent]

around the CoP moiety excludes water vapor to suppress the proton-driven irreversible oxidation of Eq. (9). Static bonding of the axial ligand with CoP in the solid state retards the process of Eq. (10). The cooperation of these effects strikingly inhibits the irreversible oxidation and prolongs the lifetime of CoP in the solid polymer membrane. In fact (Table 6.2), the lifetimes of the polymer-bound CoPs in the solid state are much longer than those of the monomeric CoPs and the CoPs in solution, and they maintain their oxygen-binding ability for over a month.[31] This enhanced stability of the carriers represents the great advantage of the solid membranes of polymer complexes.

It has been reported that a large number of transition-metal complexes coordinately bind molecular nitrogen selectively and reversibly.[34] However, these nitrogen-coordinated complexes are unstable and irreversibly undergo degradation in an open air atmosphere through dimerization of the metal complex and/or reactions with water vapor.

Cyclopentadienylcarbonylmanganese (CpMn) and benzenecarbonyl-chromium complex were introduced into polymer chains by radical copolymerization (structures *12* and *13*), which provides transparent, flexible, and solvent-free membranes.[35,36] Upon uv irradiation of the membranes in argon atmosphere, they were converted to the corresponding dicarbonyle-metals (structures *12* and *13*), which have one unsaturated coordination site to bind molecular nitrogen reversibly

$$\text{CpMn} + N_2 \rightleftharpoons \text{CpMn}-N_2. \tag{11}$$

On exposure of the structure *12* membrane to nitrogen atmosphere, the polymer membrane showed a strong ir absorption peak at 2160 cm^{-1} for $^{14}N_2$ and 2090 cm^{-1} for $^{15}N_2$, assigned to an endon-type coordinated dini-

13

[Structure 13: styrene copolymer with CONH-linked benzenechromium(CO)₂(N₂) complex bearing CH₃ substituents] trogen. The intensity reversibly increased and decreased in response to the partial pressure of nitrogen. The lifetime (half-life period) of the nitrogen-binding ability was >1 day. It has been reported that the corresponding cyclopentadienyldicarbonylnitrogenmanganese and benzenedicarbonylnitrogenchromium complexes decompose immediately after exposure to an open air. The polymer matrix protects the nitrogen complex even in air, probably because the polymer matrix retards diffusion of water vapor to the complex moiety and fixes the complex to suppress irreversible dimerization of the complex.

Sorption microbalance measurements of the membrane fragment indicated that nitrogen sorption was selectively enhanced by chemical nitrogen binding to the CpMn complex moiety and that the nitrogen-binding equilibrium curve also obeyed a typical Langmuir isotherm.[37] The equilibrium constant [K in Eq. (11)] is given in Table 6.3 with its thermodynamic parameters. K and the parameters in the cooled solution and for oxygen binding to an iron porphyrin complex are also listed in Table 6.3 as references. Comparison with the former values supported the validity of the nitrogen-binding parameters in the solid membrane state. The K value for nitrogen binding is a little smaller than that for oxygen binding. Both the enthalpy and the entropy change for nitrogen binding to CpMn are more positive than those for oxygen binding. The smaller enthalpy gain upon nitrogen coordination is compensated by a smaller entropy decrease in the immobilization of nitrogen, which provides the relatively large nitrogen-binding equilibrium constant.[37]

Rapid and reversible nitrogen binding to the CpMn moiety in the membrane was also confirmed by laser flash photolysis of the nitrogen complex.[37] The nitrogen-binding rate constants (k_{on} and k_{off}) are given also in Table 6.3 with kinetic data for the oxygen binding of the iron porphyrin. A k_{on} value of order $10^5 \, M^{-1} \, s^{-1}$ for nitrogen binding means that the organometallic nitrogen coordination is also a rapid reaction, although the rate constants are 10^2 times smaller than those for oxygen binding. This k_{on} value determined for the complex in the solid membrane state is consistent with the k_{on} value reported for the nitrogen-coordinated intermediate of the corresponding monomeric CpMn complex with time-resolved spectroscopy.[38] One of the merits of the membrane system is that kinetic and equilibrium constants of the gaseous molecule binding can be evaluated in situ spectroscopically.

The polymer–manganese complex *12* selectively binds acetylene from its mixture gas with ethane and ethylene.[39] For example, the ratio [acetylene]/[ethane] absorbed into the membrane after exposure for 10 min at 25°C to the 1:1 mixture gas of acetylene and ethane at 76 cm Hg was 4.6. A control datum for activated carbon was 0.59. ir absorption at 1740 cm^{-1} was assigned to a sideon-type coordinated acetylene to the metal ion. The acetylene coordination in response to acetylene partial pressure also obeyed Langmuir's isotherm, giving the K value for acetylene binding in Table 6.3.

Table 6.3 Rate and Equilibrium Constants and Thermodynamic Parameters for Nitrogen and Acetylene Binding to the Cyclopentadienylmanganese Complex in Sold-State Membranes

Complex	Ligand	Physical State	$10^{-2}K$ (M^{-1})	$10^{-3}k_{on}$ $(M^{-1}\,s^{-1})$	$10^{-3}k_{off}$ (s^{-1})	ΔH (kcal mol^{-1})	ΔS (e.u.)
12	THF[a]	THF soln., 0°C	8.3	61	7.4	—	—
12	N_2	solid membr., 20°C	9.8	2.9	0.3	−11	−35
12	C_2H_2	solid membr., 25°C	0.26	0.00024	0.00094	−14	−60
CpMn(CO)$_2$[b]	N_2	cyclohexane soln., 22°C	—	3.7	—	—	—
FeP[c]	O_2	toluene soln., 25°C	23	1060	46	−14	−42

[a]Tetrahydrofuran.
[b]Cyclopentadienylbiscarbonylmanganese complex.
[c]Iron (picket fence porphyrin) complex.

This K value is ~50 times smaller than that for the nitrogen binding. Although the larger enthalpy gain (negative H value in Table 6.3) suggests a stronger bond between C_2H_4 and Mn in comparison with the N_2–Mn bond, a much more negative entropy change through the acetylene binding contributes to the smaller K value. A larger entropy decrease in the coordination step for the more bulky acetylene molecule reduces the binding equilibrium constant, as compared with the coordination of the smaller nitrogen molecule. The flash photolysis procedure was also adequate to estimate k_{on} and k_{off} for acetylene binding (Table 6.3). The k_{on} value for acetylene is 10^3 times smaller than that for nitrogen binding: This indicates a more organometallic character for the acetylene coordination.

In any case, when these solid membranes of polymer complexes were placed under a gaseous pressure gradient, the complexes are expected to contribute to the permeation flux across the membranes of a specific gaseous molecule.

4. Solid Membranes for Facilitated Transport

Figure 6.1(b) schematically represents the facilitated transport of a gaseous molecule via a fixed carrier (complex) in a solid membrane. For the example of oxygen, the fixed carrier picks up oxygen specifically from air at the upstream–membrane interface. The oxygen taken up into the membrane is transferred by the fixed carriers from the upstream to the downstream side in response to the concentration gradient of oxygen across the membrane. At the membrane–downstream interface the carrier releases oxygen to the downstream side. If the passage of oxygen by fixed carriers is efficient and rapid, the oxygen transport is enhanced.

We first succeeded in the facilitated transport of a gaseous molecule through a solid membrane using kinetically active CoP membranes for selective oxygen transport. The oxygen permeability coefficient (P_{O_2}) for a solid membrane of the polymer CoP complex 7 is shown i Fig. 6.5.[24,26] P_{O_2} is larger than the nitrogen permeability coefficient (P_{N_2}) and steeply increases with a decrease in the oxygen upstream pressure [$p_2(O_2)$]. On the other hand, P_{N_2} is small and independent of the nitrogen upstream pressure [$p_2(N_2)$], because the fixed complex does not interact with nitrogen. P_{O_2} is also small and independent of $p_2(O_2)$ for the control membrane composed of the inert Co(III)P, which does not interact with oxygen. That is, the active CoP complex fixed in the membrane facilitates oxygen transport in the membrane and enhances the oxygen permeation additionally, represented as the shadowed area in Fig. 6.5. P_{O_2} increases in Fig. 6.5 with the loaded-CoP concentration in the membrane. The permeability ratios of oxygen to nitrogen (P_{O_2}/P_{N_2}) was 3.2, 5.7, and 12 for the membrane containing 0, 2.5, and 4.5% CoP complex moiety, respectively. This result indicates the possibility of high permselectivity with a polymer complex membrane.

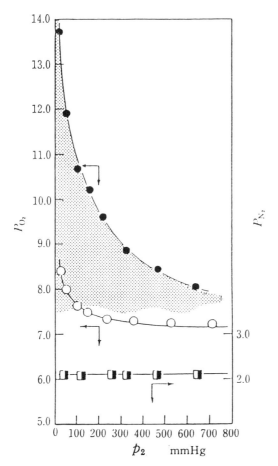

Figure 6.5 The oxygen (P_{O_2}) and nitrogen permeability coefficients (P_{N_2}) for the polymer cobaltporphyrin complex (7) membrane. The units on the ordinate are 10^{-10} cm³ (STP) cm cm⁻² s⁻¹(cm Hg⁻¹); p_2: upstream gas pressure, the complex concentration in the membrane: ● 4.5%, ○ 1.2%, at 30°C.

The time course of the permeation of gaseous molecules through membranes showed an induction period followed by permeation with a constant slope (steady state). The induction period (θ) for the oxygen permeation was longer than θ for nitrogen and was prolonged with the decrease in $p_2(O_2)$.[24–26] On the other hand, θ for nitrogen permeation was short and independent of $p_2(N_2)$. This behavior indicates that oxygen clearly interacts with the complex in the membrane and that its diffusivity in the membrane was reduced by repeated binding and dissociation of oxygen to the fixed complex. $θ_{O_2}$ was prolonged with the loaded-CoP concentration in the membrane. $θ_{O_2}$ and the $p_2(O_2)$ dependency of $θ_{O_2}$ were enhanced at lower temperature, because the oxygen-binding rate constants decreased and the equilibrium constant increased with decreasing temperature. θ is the best parameter to distinguish whether facilitation by the complex contributes to the transport or not in the polymer complex membrane.

With the increase in the loaded amount of a complex, the polymer membrane becomes too brittle for practical use under the pressure difference between up- and downstream without pinhole formation and breakage. One of the attempts succeeded in preparing a highly complex-loaded solid membrane by coating a thin (10 μm) active layer of polymer complex 7 on porous glass membrane.[40] 30% CoP-loaded membrane afforded high oxygen/nitrogen selectively of 16 even at 1 atm upstream pressure.

To improve the oxygen permselectivity and/or permeability, several polymer matrices have been applied to the facilitated-transport membrane.[41] Silicone membrane is utilized for oxygen enrichment because of its high permeability. Silicone polymer, having a large free volume, could contain a large amount of the complex while maintaining its high permeability, but oxygen transport in the silicone-bound CoP membrane (structure *14*) was barely augmented in comparison with nitrogen flux, because physical permeation is predominant in a membrane containing such highly permeable matrix polymer.[42]

Poly(1-trimethylsilyl-1-propyne) possesses the highest gas permeability of all polymer dense membranes.[43] The specific structure of this polymer with its rigid polyacetylenic backbone and the bulky and flexible trimethylsilyl substituent causes its large frozen free volume to provide an enormously high permeability (Chap. 4). The CoP complex was introduced to this frozen free volume[44]: The combination membrane showed an oxygen/nitrogen selectivity of 4.5, maintaining its high permeability ($P_{O_2} = 10^{-7}$).

14

Additionally, physical aging or a decline in the permeability with time was suppressed with added CoP. This result suggests that a combination of established membranes with the complex having gas selectivity is one effective method to give them gas permselectivity.

The upstream surface of the CoP membrane shown in structure 7 was coated with a second polymer having a smaller gas solubility.[45] The oxygen/nitrogen selectivity was increased from ~5 to 12 by the coating. The gas concentration on the upstream side of the CoP membrane is much reduced by the coated polymer, and the contribution of the facilitated transport is still greater than that of the physical permeation even under relatively high upstream oxygen pressure.

It has been reported that composite membranes containing liquid-crystal domains attain a high gas permeability.[46] Composite membrane was prepared from N-methoxybenzylidene-4-butylaniline as a liquid-crystalline phase, with polymethylmethacrylate as a support, and CoP as an oxygen carrier.[47] Although the oxygen permeability was barely increased by the CoP retained in the liquid-crystal domain, it was strongly affected by the aggregation state of the domain and was discontinuously changed at the temperatures of the crystal–nematic and nematic–isotropic phase transitions. This result suggests that the selective gaseous-molecule transport capability of the fixed complex could be controlled by the responsiveness of the polymer matrix to stimulation.

In any case, the most important advantage of the solid membrane of polymer complexes is the enhanced stability of their carrier function, as described in Section 3. For example, the CoP membrane 7 continuously worked as an oxygen-enriching membrane in laboratory air at room temperature for 6 months. The membrane of CoP coordinated to poly(fluoroalkylmethacrylate-co-vinylimidazole) was water repelling. The CoP complex in this membrane reversibly bound oxygen even under water-vapor-saturated air,[48] because the irreversible oxidation of CoP via Eq. (9) was retarded by the hydrophobic property of the polymer.

The corresponding iron porphyrin complexes in the solid membranes also bind oxygen selectively and reversibly. Although the oxygen-binding affinities of the iron porphyrin complexes were larger than those of the CoPs, the oxygen carrier lifetimes were shorter because of their lower oxidation potential.[41] Therefore, cobalt is much preferable in metal complexes as the oxygen carrier to be incorporated in membranes.

For a metal complex to function as a useful oxygen carrier in an air separation process, it must have capacity, adequate oxygen-binding thermodynamics and kinetics, acceptable cost, and chemical stability. The second notable is the family of cobalt Schiff-base complexes. Complex 5 and cobalt Schiff bases covalently bonded to polymers in the solid state were examined as oxygen adsorbents (powder) for gas chromatography and the pressure-swing method,[49,50] but their kinetic ability in oxygen binding and dissociation was much reduced after solidification.

We have reported that structure 5 coordinated to the copolymers of 4-vinylpyridine or N-vinylimidazole bind oxygen reversibly with several days' lifetime and that the polymer chain restricted irreversible oxidation through dimerization.[51] The complex with the copolymers of alkylmethacrylate gave transparent and flexible membranes.[52] The color of the membranes changed reversibly from brown under nitrogen to deep violet on exposure to oxygen, which is attributed to the oxygen-coordinated complex ($Co/O_2 = 1/1$) formation. The cobalt complex in the solid membrane binds oxygen rapidly and effectively acts as a fixed carrier for oxygen transport in the membranes.[53] The oxygen/nitrogen selectivity increased with the CoS complex concentration and was above 10 for the membrane containing 12% CoS. A high permselectivity was also measured for the CoS complex-coated porous membrane.[54]

Facilitated oxygen transport via the fixed cobalt Schiff-base complex in glassy polymer membranes has also been reported. A high oxygen/nitrogen selectivity, >10, was measured for the CoS complex membrane composed of poly(styrene-co-vinylpyridine).[55] While the glassy and rigid polymer matrix suppressed physical permeation of nitrogen and enhances selectivity, the flux through the membrane was greatly reduced to an impractical level. Drago synthesized a new cobalt Schiff-base complex covalently bonded to polystyrene (structure *15*), in which the cobalt complex was strictly immobilized to inhibit its irreversible oxidation, and studied oxygen transport through its solid membrane.[56,57] Many studies on facilitated oxygen transport in dry polymer membranes via cobalt Schiff-base complexes were recently reported using styrene–butadiene–vinylpyridine graft copolymers,[58] epoxided styrene–butadiene block copolymer,[59] and silicone.[60]

For the polymer–CoMn complex *12*, nitrogen transport through its membrane was selectively augmented due to the rapid and reversible ni-

15

trogen-binding to the fixed CpMn complex. Nitrogen permeability coefficient (P_{N_2}) increased with a decrease in $p_2(N_2)$, while oxygen permeability was independent of $p_2(O_2)$ (Fig. 6.6). The CpMn complex in the membrane interacted specifically with nitrogen and not with oxygen, and nitrogen transport through the membrane was facilitated by the CpMn complex.

The acetylene permeability coefficient ($P_{C_2H_2}$) was also plotted in Fig. 6.6, but $P_{C_2H_2}$ enhancement and $p_2(C_2H_2)$ dependency of $P_{C_2H_2}$ were not observed; the acetylene transport through the membrane was not facilitated. That is, acetylene is totally immobilized to the CpMn complex in the polymer. In Table 6.3, the acetylene-binding rate constants, especially the k_{off} value, were much smaller than those of the oxygen and nitrogen binding. This kinetically inactive binding of acetylene hardly contributes to acetylene transport in the polymer complex membrane.

The binding reaction character of the fixed complex with a gaseous molecule clearly relates to the selective transport of the gaseous molecule across the solid membrane.

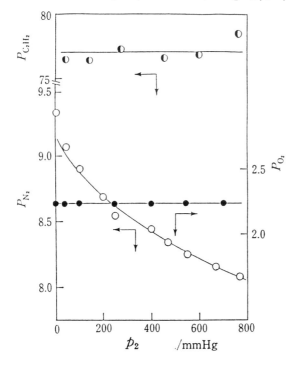

Figure 6.6 Permeability coefficients of oxygen (P_{O_2}), nitrogen (P_{N_2}), and acetylene ($P_{C_2H_2}$) for the polymer manganese complex membrane (structure 12). The ordinate units are 10^{-10} cm^3 (STP) cm cm^{-2} s^{-1}(cm Hg^{-1}); p_2: upstream gas pressure at 40°C.

5. Transport Mechanism and the Effects of Fixed Complexes

The facilitated transport of gaseous molecules in polymer complex membranes is considered to occur by a dual-mode transport mechanism, which was first proposed to explain the transport behavior of condensable gases such as carbon dioxide in glassy polymers.[61–63] A frozen free volume or microvoid that arises from the nonequilibrium nature of glassy polymers acts as a Langmuir sorption or tapping site for the condensable gases and enhances their permeation. The CoP polymer membrane also sorbed oxygen by a dual-mode process, as described in Section 3: physical sorption to the polymer domain according to Henry's law and additional chemical sorption to the CoP complex according to a Langmuir isotherm. The sorbed oxygen interacted with the fixed CoP complex rapidly and reversibly and was not immobilized by it during passage through the membrane, as shown in Section 4. Thus the oxygen transport was accelerated by the additional Langmuir mode in addition to the Henry mode. Figure 6.7 schematically represents this oxygen permeability in the fixed carrier membrane that is governed by both the Henry and Langmuir modes; that is, oxygen physically dissolves in and diffuses through the polymer matrix via the upper permeation route. In addition, the lower permeation route represents that oxygen is specifically and chemically taken up into the membrane by the selective binding reaction to the CoP complex and diffuses through the fixed CoP complex moiety by repeating the binding and dissociation reaction to and from the complex. The dual-mode transport model is mathematically given as[61–64]

$$P = k_D D_{DD} + \frac{C'_C K D_{CC}}{(1 + Kp_2)} + \frac{C'_C K D_{CD} - k_D D_{DC}}{(1 + Kp_2)}$$
$$+ \frac{2k_D D_{DC}}{Kp_2} \ln(1 + Kp_2). \tag{12}$$

Here k_D is the physical solubility coefficient for the Henry's-law mode. D_{DD}, D_{DC}, D_{CD}, and D_{CC} are the diffusion coefficients for the Henry's-law physical permeation, for the diffusion from the polymer matrix to the fixed carrier (complex), for the diffusion from the fixed carrier to the polymer matrix, and for hopping between the fixed carriers, respectively. C'_C is the fixed carrier concentration in the membrane. The total permeability coefficient (P) is equal to the sum of the first term, which represents the Henry's-law mode attributed to a physical permeation through a polymer matrix and is given by the product of the solubility and the diffusivity, the second term, which represents the Langmuir mode attributed to a specific binding and diffusion of oxygen to and via the fixed carrier (complex) and is given by the product of the Langmuir isotherm and the diffusivity via

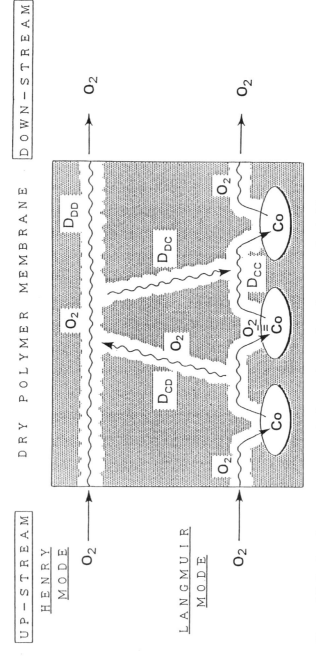

Figure 6.7 Dual-mode oxygen transpoart in the membrane of a polymer complex (solid membrane containing fixed carrier).

the fixed carrier, and the third and fourth term, which are also the Langmuir mode attributed to the exchange between the first and second terms.

The experimental data in Fig. 6.5 were analyzed with data on the $p_2(O_2)$ dependency of the time lag (θ) by the dual-mode transport model and gave the parameters listed in Table 6.4. D_D and k_D values agreed with each other for the CoP membranes containing structures 7–10 and with those determined for the inert Co(III)P membrane. This result means that the polymer matrix is not influenced by CoP species and supports the calculation procedure based on Eq. (12). The P_{O_2} versus $p_2(O_2)$ curve was calculated with the parameters in Table 6.4 and drawn as solid lines in Fig. 6.4.[28,29] The experimental plots agree with the solid lines, which also support the facilitated and dual-mode transport of oxygen in the membrane and a pathway of oxygen permeation via the fixed complex (carrier).

The oxygen diffusion coefficients in Table 6.4 suggest the following[26,28,29]:

1. The four diffusion coefficients are in the order of $D_{DD} > D_{DC} > D_{CD} > D_{CC}$, which is in accordance with their definition, for example, the contribution of the reaction step to and from the carrier: $k_{on}/k_{off} \sim D_{DC}/D_{CD}$.
2. The ratio D_{CC}/D_{DD} shows that the diffusion constant, postulated in Fig. 6.7 for the hopping pathway of oxygen between the fixed carriers, is approximately one-tenth that of the physical permeation in the membrane.
3. The reaction rate constants, k_{on} and k_{off}, determined spectroscopically for oxygen binding and dissociation to and from the CoP as a fixed oxygen carrier given in Table 6.1, are clearly related to the diffusion coefficients D_{DC}, D_{CD}, and D_{CC} determined from the transport measurement. Especially D_{CC} was directly related to k_{off}.
4. The CoP carrier, structure 10, with larger k_{on} and k_{off}, gives a larger D_{DC}, D_{CD}, and D_{CC} and brings about highly efficient facilitation in the oxygen transport with an oxygen/nitrogen selectivity of >20.

Table 6.4 Diffusion Constants and their Activation Energies (ΔE) for Facilitated Oxygen Transport in Polymer Cobalt Porphyrin Complex Membranes[a]

Complex	$10^6 D_{DD}$	$10^7 D_{DC}$	$10^8 D_{CD}$	$10^9 D_{CC}$	ΔE_{DD}	ΔE_{DC}	ΔE_{CD}	ΔE_{CC}
	(cm² s⁻¹)				(kcal mol⁻¹)			
7	2.2	0.97	0.86	3.1	11	14	16	21
8	2.2	2.2	1.5	7.3	10	14	16	19
9	2.2	2.6	2.0	9.0	11	14	16	19
10	2.2	3.2	22	144	10	13	15	18

[a]k_D[cm³ (STP) cm⁻³ (cm Hg⁻¹)] = 7.2 ± 0.2; C_C' [cm³ (STP) cm⁻³] = 0.2 ± 0.02.

Table 6.5 Diffusion Constants and Their Activation Energies (ΔE) for Facilitated Oxygen Transport in the Polymer Membranes Containing Cobaltporphyrin Complex 7

Alkyl Chain of 7	T_g^a (°C)	$10^6 D_{DD}$	$10^7 D_{DC}$	$10^8 D_{CD}$	$10^9 D_{CC}$	ΔE_{DD}	ΔE_{DC}	ΔE_{CD}	ΔE_{CC}
		(cm² s⁻¹)				(kcal mol⁻¹)			
butyl	27	1.6	1.9	1.3	7.9	12	15	17	20
octyl	−3	2.4	2.2	1.5	7.9	11	14	16	20
lauryl	−45	6.2	2.9	2.6	7.8	10	14	16	20

$^a T_g$: glass transition temperature; C_C' [cm³ (STP) cm⁻³] = 0.2 ± 0.02.

5. The apparent activation energies for the four diffusion steps show the same tendency, e.g., E_{DD} and E_{CC} are independent and variable of the carrier (complex) species, respectively.

On the other hand, Table 6.5 indicates the effects of polymer matrix on the four diffusion coefficients.[65] The glass transition temperature of the membrane decreases, and the free volume of the polymer matrix is increased, with the alkyl chain length of the copolymer ligand coordinated to the complex. In response to this free-volume increase in the membrane, the diffusion constant D_{DD} through the polymer matrix is augmented, and its activation energy is reduced. In contrast, the postulated hopping diffusion step between the carriers (D_{CC}) is hardly affected by this property of the polymer matrix.

The oxygen diffusion coefficient via the carrier (complex) fixed in the polymer membrane (D_{CC}) is plotted against the dissociation rate constant of the coordinated oxygen from the complex for the membrane composed of various cobalt porphyrin complexes in Fig. 6.8.[66] The logarithmic plots of D_{CC} versus k_{off} show a linear relationship. D_{CC} increases with k_{off} and is independent of K. It is first concluded that the dissociation kinetic constant of the permeate gaseous molecule from the fixed complex is clearly related to the permeate diffusion coefficient via the fixed complex in the

Figure 6.8 Correlation of the dissociation rate constant of bound oxygen (k_{off}) with diffusion coefficient via the fixed complex (D_{CC}) for polymer cobalt porphyrin membranes. Circle: k_{off}, square: K.

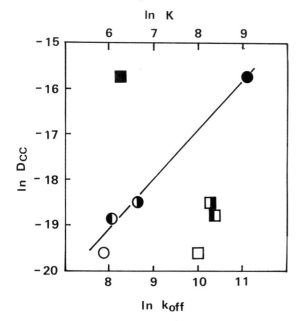

membrane. The increase in D_{CC} directly yields the enhancement of P in Eq. (12).

Second, Eq. (12) suggests that an increase in the oxygen-binding equilibrium constant (K) also yields enhanced facilitation. P_{O_2} was larger for the membrane of the CoP coordinated to the imidazolyl residue such as complex 7, which has a larger K value ($P_{O_2} = 2.2 \times 10^{-9}, K = 5.6 \times 10^{-2}$ cm Hg^{-1}), in comparison with the CoP coordinated to the pyridyl residue ($P_{O_2} = 1.0 \times 10^{-9}, K = 1.4 \times 10^{-2}$).[66] It is concluded that the carrier not only binds the permeate molecule specifically and strongly (large K), but also releases (large k_{off}) and hands it over quickly (large D_{CC}), as can be schematically understood in Fig. 6.1(b), in order to realize facilitated transport in the solid membrane.

The variation in the binding capability of the complex often increases its binding affinity but is accompanied with a decrease in the kinetic parameters. That is, K is k_{on} divided by k_{off} [Eq. (6)], and the increase in the dissociation rate constant sacrifices the binding affinity. But, for example, if one could synthesize a complex with both a stronger binding affinity and kinetic parameter (around only twice both K and k_{off} of the present complex such as 10) of $K = 1$ cm Hg^{-1} and a larger kinetic parameter of $D_{CC}/D_{DD} = 0.2$, an outstanding permeability coefficient of $P = 10^{-6}$ cm^3 (STP) cm^{-2} s^{-1} (cm Hg^{-1}) and an oxygen-nitrogen selectivity of >100 are estimated with Eq. (12) for $p_2 = 15$ cm Hg (air), $D_{DD} = 1 \times 10^{-6}$ cm^2 s^{-1}, $k_D = 1 \times 10^{-3}$ cm^3 (STP) cm^{-3} (cm Hg^{-1}), and $C'_C = 10$ cm^3 (STP) cm^{-3}. The facilitated-transport membranes of polymer complexes have a strong potential as next-generation permselective membranes.

6. Conclusion and Further Studies

Figure 6.9 shows plots of the oxygen permeability coefficient P_{O_2} versus oxygen/nitrogen permselectivity for the membranes of currently available polymers.[67] Although the effects of the chemical structure of polymers on their gas permeability are being studied, as described in Chapter 4, the P versus selectivity relationship for the oxygen/nitrogen separation still remains in an inverse correlation, as can be seen in Fig. 6.9. The membranes based on the facilitated-transport oxygen-carrier (complex) chemistry described here can exhibit both a significantly higher oxygen selectivity and a higher oxygen permeability than those of present-day polymers or passive membranes: Their closed plots surpass inverse correlation. The membranes described here can produce $> 80\%$ oxygen or 99.9% nitrogen from air under certain conditions.

A number of future subjects remain to improve the polymer complex membranes up to the level of a practical facilitated-transport membrane system. To increase the permeation flux through the membrane it must be thin, but the facilitation efficiency is often a function of the membrane thickness, which implies that there is sometimes a kinetic limitation to the

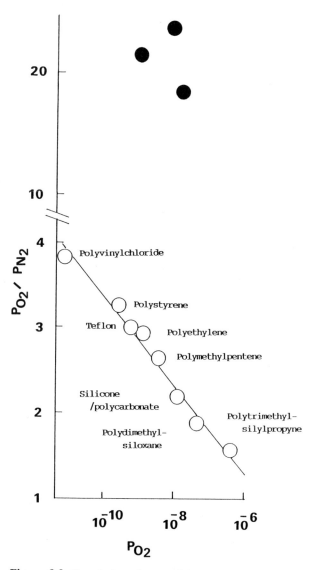

Figure 6.9 Correlation of oxygen/nitrogen permselectivity (P_{O_2}/P_{N_2}) with oxygen permeability (P_{O_2}) for membranes. Filled symbols: polymer cobalt porphyin membranes; abscissa units: cm^3 (STP) cm cm^{-2} s^{-1}(cm Hg^{-1}).

facilitated transport. At very short diffusional path lengths in a thin membrane, the facilitated transport may be partially limited by the reaction kinetics of the complex (carrier). It appears that a complex having an approximate binding affinity with a much faster off rate is needed to be effective with very thin membranes. Much more research is also needed to elucidate the chemistry of the solid-state carriers, such as in situ structure characterization of the fixed complexes and their chemical reactivity in the solid state.

Hemoglobin binds oxygen not only selectively and reversibly but also efficiently according to an allosteric mechanism: The oxygen-binding equilibrium curve becomes sygmoidal, as represented in Fig. 6.10. The transport efficiency between the up- and downstream sides is greatly enlarged by this S-like curve. Synthetic metal complexes with pseudoallosteric binding functions have recently been reported.[68] The allosteric complexes may provide new possibilities for facilitated-transport membranes.

The permselectivity of the facilitated-transport membranes was directly related to the concentration of the carrier complex in the membranes. The active fixed carrier concentration was estimated by both gas sorption tests into the membrane and gas permeation tests through the membrane, as described in Sections 3 and 4, but the effective carrier concentration determined by the former equilibrium method was often larger than that by the latter kinetic estimation. This result suggests that a part of the fixed carrier does not contribute to the facilitated transport flux probably because it is fixed in an unfavorable orientation to the flux. The face of the carrier's binding site is to be orientationally immobilized to form a continuously hopping pathway of the facilitated molecule. The molecular assembly and liquid-crystal orientation of the carrier molecules[69] may realize such an alignment of the carrier.

A variety of microporous polymer, ceramic, glass, and metal membranes has been developed in recent years, and gas permeation through these microporous membranes has achieved different transport mechanisms, such as surface diffusion on the pore walls. A microporous glass membrane, whose pore walls were modified with a CoP complex, was preliminarily reported[70]: Oxygen selectively diffuses via the complex-modified surface

Figure 6.10 A carrier (complex) with allosteric binding ability. Solid line: sorption isotherm for a carrier with allosteric (S-like) binding; dashed line: sorption isotherm for a carrier with Langmuir-type binding; p: pressure; p_1: downstream pressure; p_2: upstream pressure; B (%): binding degree of the carrier.

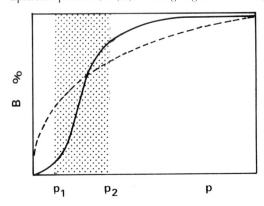

of the pore walls (pore size ~40 Å), and the membrane could prove useful for the separation of oxygen from air with a selectivity of ~3 (for a control glass membrane ~1) and a high permeability of 10^{-6} cm^3 (STP) cm cm^{-2} s^{-1} (cm Hg^{-1}). Selective surface diffusion of gaseous molecules may become a challenge by suitable modification of the pore walls of a microporous membrane with complexes.

Specific combinations of binding reactions between gaseous molecules and metal complexes have now been established in the field of inorganic chemistry. The introduction of these metal complexes to polymers offer the possibility of membranes for highly selective and efficient gas separation. The practical advantages of the ease of membrane formation, membrane and carrier stabilities, and the feasibility of polymer complex membranes provide a great incentive for the research effort on this type of facilitated-transport membrane.

REFERENCES

1. E. Tsuchida and H. Nishide, *Adv. Poly. Sci.* **24**, 1 (1977).
2. E. Tsuchida and H. Nishide, *Topics Curr. Chem.* **132**, 64 (1986).
3. *Macromolecular Complexes: Dynamic Interactions and Electronic Processes*, E. Tsuchida, Ed., VCH Publishers, New York (1991).
4. P. F. Scholander, *Science* **131**, 585 (1959).
5. W. J. Ward and W. L. Robb, *Science* **156**, 1481 (1970).
6. E. Hemmingsen and P. F. Scholander, *Science* **132**, 1379 (1960).
7. M. Mochizuki and R. E. Forster, *Science* **138**, 897 (1962).
8. R. J. Bassett and J. S. Schultz, *Biochim. Biophys. Acta* **211**, 194 (1970).
9. Bend Research Inc., *U.S. Pat.* 4,609,383 (1986).
10. B. M. Johnson, R. W. Baker, S. L. Matson, K. L. Smith, E. C. Roman, M. E. Tuttle, and H. K. Lonsdale, *J. Membr. Sci.* **31**, 31 (1987).
11. J. C. Stevens and D. H. Busch, *J. Amer. Chem. Soc.* **102**, 3285 (1980).
12. S. L. Matson and H. K. Lonsdale, *J. Membr. Sci.* **31**, 69 (1987).
13. I. R. Bellobono, F. Muffato, E. Selli, L. Righetto, and R. Tacchi, *Gas Sep. Pur.* **1**, 103 (1987).
14. J. A. T. Norman, G. P. Pez, and D. A. Roberts, *Proc. Annu. IUCCP Symp.*, 107 (1987).
15. Y. Kaboyashi, I. Konno, and J. Matsuura, *Proc. Macromol. Metal Complexes Symp.*, 345 (1986).
16. K. Okamoto, K. Yamada, and S. Yamanouchi, *Proc. ICOM*, 563 (1990).
17. C. A. Koval, R. D. Noble, J. D. Way, B. Louie, E. Reyes, B. R. Bateman, G. M. Horn, and D. L. Reed, *Inorg. Chem.* **24**, 1147 (1985).
18. J. Matsuura, T. Ohuchi, and M. Okada, *Proc. ICOM*, 781 (1987).
19. D. V. Laciak and G. P. Pez, *Sep. Sci. Tech.* **25**, 1295 (1990).
20. D. V. Laciak, G. P. Pez, and P. M. Burban, *Proc. NAMS Ann. Met.*, 197 (1988).

21. O. H. LeBlanc, W. J. Word, S. L. Matson, and S. G. Kimura, *J. Membr. Sci.* **6**, 339 (1980).
22. M. Teramoto, H. Matsuyama, T. Yamashiro, and Y. Katagama, *J. Chem. Eng. Jpn.* **19**, 419 (1986).
23. T. Spontarelli, C. A. Koval, and R. D. Noble, *Proc. ICOM*, 649 (1990).
24. H. Nishide, M. Ohyanagi, O. Okada, and E. Tsuchida, *Macromolecules* **19**, 494 (1986).
25. H. Nishide, M. Ohyanagi, O. Okada, and E. Tsuchida, *Macromolecules* **21**, 2910 (1988).
26. H. Nishide, H. Kawakami, T. Suzuki, T. Azechi, and E. Tsuchida, *Macromolecules* **23**, 3714 (1990).
27. H. Nishide, H. Kawakami, S. Toda, E. Tsuchida, and Y. Kamiya, *Macromolecules* **24**, 5851 (1991).
28. H. Nishide, M. Ohyanagi, O. Okada, and E. Tsuchida, *Macromolecules* **20**, 417 (1987).
29. H. Nishide, H. Kawakami, Y. Soejima, and E. Tsuchida, *Macromolecules* **25**, in press.
30. E. Antonini and M. Brunori, *Hemoglobin and Myoglobin in their Reaction with Ligands*, North-Holland, Amsterdam (1971).
31. H. Nishide, H. Kawakami, W. Inoue, and E. Tsuchida, *Macromolecules* **24**, in press.
32. R. D. Jones, D. A. Summerville, and F. Basolo, *Chem. Rev.* **79**, 139 (1979).
33. J. P. Collman, *Acc. Chem. Res.* **10**, 265 (1977).
34. R. A. Henderson, G. F. Leigh, and C. J. Piekett, *Adv. Inorg. Chem.* **27**, 198 (1983).
35. Y. Kurimura, F. Ohta, H. Nishide, and E. Tsuchida, *Makromol. Chem.* **183**, 2889 (1982).
36. H. Nishide, H. Kawakami, E. Tsuchida, and T. Kurimura, *J. Macromol. Sci.-Chem.* **A24**, 1339 (1988).
37. H. Nishide, H. Kawakami, Y. Kurimura, and E. Tsuchida, *J. Amer. Chem. Soc.* **111**, 7175 (1989).
38. B. S. Creaver, A. J. Dixon, and M. Polialkoff, *Organometallics* **6**, 2600 (1987).
39. H. Kawakami, K. Tsuda, H. Nishide, and E. Tsuchida, *Macromolecules* **24**, 3310 (1991).
40. H. Ohno and M. Ikeda, *Proc. ICOM*, 667 (1990).
41. M. Ohyanagi, H. Nishide, K. Suenaga, and E. Tsuchida, *Macromolecules* **21**, 1590 (1988).
42. H. Nishide, M. Ohyanagi, K. Suenaga, and E. Tsuchida, *J. Polymer Sci., Polym. Chem.* **29**, 1439 (1989).
43. T. Masuda, E. Isobe, T. Higashimura, and K. Takada, *J. Amer. Chem. Soc.* **105**, 7473 (1983).
44. H. Nishide, H. Kawakami, Y. Sasame, and E. Tsuchida, *J. Polymer Sci., Polym. Chem.* **32**, in press.
45. H. Nishide, M. Ohyanagi, Y. Funada, T. Ikeda, and E. Tsuchida, *Macromolecules* **20**, 2321 (1987).

46. T. Kagyama, S. Washizu, and M. Takayanagi, *J. Appl. Polymer Sci.* **29**, 3955 (1984).
47. H. Kawakami, K. Tsuda, H. Nishide, and E. Tsuchida, *Polymers Adv. Tech.* **2**, 2323 (1991).
48. H. Nishide, H. Kawakami, W. Inoue, and E. Tsuchida, *Macromolecules* **24**, in press.
49. J. N. Gillis and R. E. Sievers, *Anal. Chem.* **57**, 1572 (1985).
50. D. Woehrle and H. Bohlen, *Makromol. Chem.* **187**, 2081 (1986).
51. E. Tsuchida, *J. Macromol. Sci.-Chem.* **A13**, 545 (1979).
52. H. Nishide, M. Ohyanagi, H. Kawakami, and E. Tsuchida, *Bull. Chem. Soc. Jpn.* **59**, 3213 (1986).
53. E. Tsuchida, H. Nishide, M. Ohyanagai, and H. Kawakami, *Macromolecules* **20**, 1907 (1987).
54. K. Sugie, *ACS Polymer Mater. Sci. Eng.*, 59 (1987).
55. M. S. Delaney, D. Reddy, and R. A. Wessling, *Proc. NAMS Ann. Met.*, 49 (1988).
56. R. S. Drago and K. J. Balkus, *Inorg. Chem.* **25**, 716 (1986).
57. M. J. Barnes, R. S. Drago, and K. J. Balkus, *J. Amer. Chem. Soc.* **110**, 6780 (1988).
58. J. M. Yang and G. H. Hsiue, *J. App. Polymer Sci.* **39**, 1475 (1990).
59. G. H. Hsiue, J. M. Yang, P. S. Lee, and C. Y. Liaw, Jr., *J. Polymer Sci. Polymer Chem.* **28**, 3363 (1990).
60. J. Sakai, H. Takenaka, and E. Torikai, *Jpn. J. Membr. Sci.* **31**, 227 (1987).
61. D. P. Paul and W. J. Koros, *J. Polymer Sci., Polymer Phys.* **14**, 675 (1976).
62. J. H. Petropoulos, *J. Polymer Sci., Polymer Phys.* **8**, 1797 (1970).
63. R. M. Barrer, *J. Membr. Sci.* **18**, 25 (1984).
64. G. H. Fredrickson and E. Helfand, *Macromolecules* **18**, 2201 (1985).
65. H. Nishide, H. Kawakami, T. Suzuki, Y. Soejima, and E. Tsuchida, *Macromolecules* **24**, 6306 (1991).
66. E. Tsuchida, H. Nishide, M. Ohyanagai, and O. Okada, *J. Phys. Chem.* **92**, 6461 (1988).
67. *Polymer Handbook*, John Wiley, New York (1975).
68. E. Tsuchida, S. G. Wang, M. Yuasa, and H. Nishide, *J. Chem.Soc. Chem. Commun.*, 179 (1986).
69. H. Nishide, M. Yuasa, T. Hashimoto, and E. Tsuchida, *Macromolecules* **20**, 459 (1987).
70. H. Nishide, H. Kawakami, E. Tsuchida, and H. Tsujikawa, *Proc. ACS Ann. Met.*, 1234 (1991).

7
Polymer Complex for the Separation of Carbon Monoxide and Ethylene

Hidefumi Hirai

1. Introduction
2. Absorbents for CO Separation
   2.1. Metal Complex Solution
   2.2. Polystyrene–$AlCuCl_4$ Complex Solution
   2.3. Complex Formation of Polystyrene with Metal Complex
3. Absorbents for Ethylene Separation
   3.1. Polystyrene–$AlCuCl_4$ Complex Solution
   3.2. Polystyrene–$AgAlCl_4$ Complex Solution
4. Adsorbents for CO Separation
   4.1. Polystyrene–$AlCuCl_4$ Complex Solid
   4.2. Polystyrene with Amino-Group–CuCl Complex Solid
5. Adsorbents for Ethylene Separation
   5.1. Polystyrene–$AlCuCl_4$ Complex Solid
   5.2. Polystyrene–$AgAlCl_4$ Complex Solid
   5.3. Polystyrene with Amino-Group–CuCl Complex Solid

1. Introduction

Carbon monoxide (CO) is produced by the partial oxidation of hydrocarbons, the water–gas reaction, or steam hydrocarbon reforming. In addition, large amounts of CO are contained in byproduct gases from the iron-making industry. All of these sources of CO are gas mixtures with hydrogen, nitrogen, carbon dioxide, methane, water, and so on.

Ethylene is produced by the thermal cracking of naphtha, natural gas, and petroleum refinery gas. Furthermore, considerable amounts of ethylene are present in coke-oven gas and byproduct gas from the fluid catalytic cracking process. In most cases, ethylene is obtained in gas mixtures with methane, ethane, propylene, propane, hydrogen, nitrogen, carbon dioxide, carbon monoxide, water, etc.

Therefore, technical improvements in the separation of carbon monoxide and ethylene are necessary for their effective utilization. The cryogenic separation process, using the differences in boiling points of the gases, is well known as a separation method for carbon monoxide and ethylene.[1] This process, however, requires large amounts of energy to achieve the necessary extremely low temperatures, and requires the complete removal of carbon dioxide and water, which can cause blockage in the refrigeration unit. The molecular weights of carbon monoxide, ethylene, and nitrogen are 28.01, 28.05, and 28.01, respectively. The boiling points of carbon monoxide and nitrogen are -191.5 and $-195.8°C$, respectively. Consequently, the separation of these gases by physical methods is difficult, especially the separation of carbon monoxide from the mixture with nitrogen.

The selective separation of carbon monoxide and ethylene is achieved by selective complex formation. Several metal complexes can form π complexes with carbon monoxide and ethylene. The metal complex solutions absorb carbon monoxide and ethylene selectively by π-complex formation at ambient temperature and release the absorbed gas at a higher temperature by decomplexation by the thermal swing method. The solids of polymer–metal complexes adsorb carbon monoxide and ethylene selectively under atmospheric pressure by π-complex formation, and release the adsorbed gas under reduced pressure by decomplexation with the pressure swing method.

Most metal complexes are easily deactivated by reaction with other components, especially with water, in the gas mixture by disproportionation and by association with themselves. The polymers in the polymer–metal complexes play an important part in protecting and supporting the metal complexes.

2. Absorbents for CO Separation

2.1. Metal Complex Solutions

The copper–ammonium–salt process has retained an important position for many years for the removal of small amounts of carbon monoxide in the synthesis–gas mixture after the shift–conversion reaction.[1] In this process, an aqueous solution of copper–ammonium formate, carbonate, or acetate absorbs carbon monoxide at higher pressures than 100 atm. The π coordination of carbon monoxide occurs on the cuprous ion in the complex solution, which is unstable owing to reduction, oxidation, and disproportionation of the cuprous ion. In addition, this solution absorbs carbon dioxide.

Recently, a separating process using a toluene solution of aluminum copper (I) chloride ($AlCuCl_4$) has been proposed.[2,3] With this absorbent, CO is separated under mild conditions, and the absorbent is stable against oxygen and carbon dioxide.[2,3] However, the CO absorbing capacity of the absorbent rapidly and irreversibly decreases on contact with the gas mixture containing water, since $AlCuCl_4$ vigorously reacts with water as follows[4]:

$$2AlCuCl_4 + H_2O \rightarrow AlCuCl_4\, Al(OH)Cl_2 + CuCl + HCl \quad (1)$$

$$AlCuCl_4\, Al(OH)Cl_2 \rightarrow AlCuCl_4\, AlOCl + HCl. \quad (2)$$

Thus the water content of gas mixtxures must be reduced to less than 1 ppm prior to CO separation with this absorbent.[2]

The $AlCuCl_4$ complex solution composed of copper (I) chloride (20 mmol), aluminum chloride (20 mmol), and toluene (20 ml) absorbs carbon monoxide rapidly from 0.8:0.2 CO/nitrogen mixture under 1 atm at 20°C.[5] The molar ratio of absorbed CO to the admitted copper (I) chloride reaches 0.70 in 30 min and 0.85 in 120 min. The absorbed CO is released in 5 min at 90°C:

$$AlCuCl_4(solvent) + CO \underset{\Delta}{\rightleftharpoons} AlCuCl_4(CO) + solvent. \quad (3)$$

After contact with 5L nitrogen containing 2 mmol water [corresponding to 10 mol % of the admitted copper (I) chloride], the CO absorbing capacity decreases by 20%, according to Eqs. (1) and (2).

2.2. Polystyrene–$AlCuCl_4$ Complex Solution

A polystyrene–$AlCuCl_4$ complex solution is prepared by incubating copper (I) chloride (20 mmol), aluminum chloride (20 mmol), and polystyrene (number-averaged degree of polymerization 420; 20 meq in styryl residue) in toluene (20 mL) at room temperature to 50°C for 4 h under dry nitrogen. The resulting polymer complex solution exhibits almost the same CO absorbing and releasing capacities as those of the $AlCuCl_4$ complex solution without polystyrene, as shown in Fig. 7.1.[5]

Both the rate of absorption and the equilibrium value of the absorbed CO after contact with nitrogen gas containing water (10 mol % to admitted CuCl) are identical with those prior to the contact. Thus the water shows no measurable deactivation effect on the polystyrene–$AlCuCl_4$ complex solution.

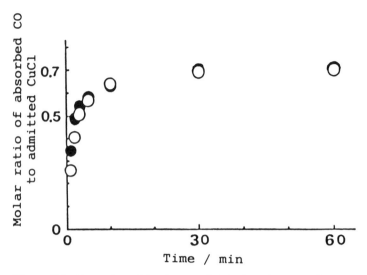

Figure 7.1 Absorption of CO from an 8:2 CO/N_2 mixture under 1 atm at 20°C by a toluene solution of poilystyrene–$AlCuCl_4$: before (○) and after (●) the contact of the complex solution with 10 mol % water to admitted copper (I) chloride. The initial volume of the gas mixture is 1500 mL.

2.3. Complex Formation of Polystyrene with Metal Complex

The equimolar mixture of copper (I) chloride and aluminum chloride becomes homogeneous on being stirred in toluene at 50°C for 10–30 min, indicating formation of $AlCuCl_4$. The $AlCuCl_4$ molecule in toluene has been proposed to have a structure with two chlorine bridges between one aluminum atom and one copper atom, based on the result of ^{27}Al-NMR spectroscopy.[6]

The toluene solution of $AlCuCl_4$ and polystyrene exhibits a strong charge-transfer band in the 380–500 nm region. The charge-transfer band has two peaks at 370–380 and 460 nm. In contrast, benzene and toluene solutions of $AlCuCl_4$ without polystyrene show a very weak charge-transfer band around 370–380 nm, and no band is observed at wavelengths longer than 400 nm.[6]

A 1,3-diphenylpropane solution of the $AlCuCl_4$ complex exhibits a strong absorption band similar to that of the polystyrene–$AlCuCl_4$ solution.[7] On the other hand, both 1,2-diphenylethane and 1,4-diphenylbutane solutions of $AlCuCl_4$ do not have such strong charge-transfer bands. Consequently, the strong charge-transfer interaction with $AlCuCl_4$ requires two phenyl groups connected by a linkage of three methylenes. Thus 1,3-diphenylpropane is found to be a suitable monomer model for the styrene in the polymer–metal complex.

The chemical shift of the carbon atoms of 1,3-diphenylpropane in 1,2-

dichloroethane changes on addition of AlCuCl$_4$.[6] The chemical shifts of phenyl carbons at the meta and para positions move to higher magnetic fields, while those of other carbons of phenyl group and methylene groups move toward lower magnetic fields.

A continuous variation plot using the chemical-shift changes of the meta and para carbons clearly indicates a 1:1 interaction between 1,3-diphenylpropane and AlCuCl$_4$ in 1,2-dichloroethane.[8]

The equilibrium constant (K) and enthalpy change ($-\Delta H$) for complex formation between 1,3-diphenylpropane and AlCuCl$_4$ are much larger than those for formation between toluene and AlCuCl$_4$: K at 273 K = 167 L/mol, $-\Delta H$ = 16.3 kJ/mol for the 1,3-diphenylpropane complex; K at 273 K = 3.8 L/mol, $-\Delta H$ = 5.5 kJ/mol for the toluene complex.[8] Thus, the binding energy between AlCuCl$_4$ and 1,3-diphenylpropane is much larger than that between AlCuCl$_4$ and toluene.

The 1,3-diphenylpropane–AlCuCl$_4$ complex solution absorbs carbon monoxide selectively and is not deactivated by the water contained in the gas.[9]

All these results indicate that a two-way interaction of 1,3-diphenylpropane with AlCuCl$_4$ takes place through π coordination of two phenyl groups. The two-way interaction should be much stronger than a one-way interaction through one phenyl group.

The proposed structure of the polystyrene–AlCuCl$_4$ complex is shown in Fig. 7.2.[8] The two-way interaction between the adjacent phenyl groups

Figure 7.2 Complex formation of aluminum copper (I) chloride with polystyrene and with carbon monoxide.

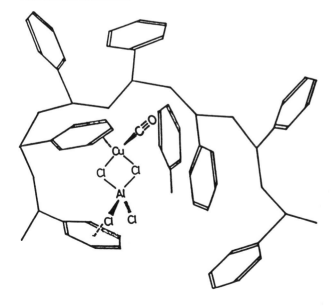

of polystyrene and AlCuCl$_4$ makes the complex formation effective, and the hydrophobic polystyrene chain may inhibit the approach of water to the metal complex. The toluene molecule acts as a solvent and has a one-way interaction with the AlCuCl$_4$ complex. The toluene molecule can be replaced by a CO molecule.

3. Absorbents for Ethylene Separation

3.1. Polystyrene–AlCuCl$_4$ Complex Solution

The polystyrene–AlCuCl$_4$ complex solution absorbs ethylene from a 0.94:0.06 ethylene/nitrogen mixture under 1 atm at 20°C more rapidly than CO from a 0.8:0.2 CO/nitrogen mixture. The molar ratio of absorbed ethylene to admitted copper (I) chloride reaches 1.6 in 30 min and 1.80 in 120 min.[10] The absorbed ethylene is released by an elevation of the temperature from 20 to 100°C. The π coordination of ethylene to the Cu(I) ion in the complex solution is stronger than that of CO to the Cu(I) ion.

The Friedel–Crafts reaction between the absorbed ethylene and toluene used as a solvent takes place when the temperature is raised, producing ethyltoluene. The reaction rate depends on the molar ratio of copper (I) chloride to aluminum chloride in the feed. At a molar ratio of 1.0, 14% of the absorbed ethylene reacts with toluene in each absorption–release cycle. At a molar ratio of 1.1, the reaction is suppressed and 1.9% of the absorbed ethylene reacts with toluene.[10]

3.2. Polystyrene–AgAlCl$_4$ Complex Solution

A polystyrene–aluminum silver chloride (AgAlCl$_4$) complex solution is prepared by incubating silver chloride, aluminum chloride, and polystyrene in toluene at room temperature to 50°C for 4 h under dry nitrogen, in a similar manner to a polystyrene–AlCuCl$_4$ complex solution.[10,11]

The resulting polymer complex solution absorbs ethylene moderately from a 0.94:0.06 ethylene/nitrogen mixture under 1 atm at 20°C. The molar ratio of absorbed ethylene to the admitted AgCl reaches 0.50 in 30 min and 0.54 in 120 min. The absorbed ethylene is released almost completely by an elevation of the temperature from 20 to 100°C.

The polystyrene–AgAlCl$_4$ complex solution does not absorb CO in a measurable amount from a 0.94:0.06 CO-nitrogen mixture. Thus the selectivity of ethylene against CO for the polystyrene–AgAlCl$_4$ complex solution is much greater than that for the polystyrene–AlCuCl$_4$ complex solution, which has a selectivity of 2.1.[10,11]

4. Adsorbents for CO Separation

4.1. Polystyrene–AlCuCl$_4$ Complex Solid

Solid CO adsorbents are prepared using a cross-linked polystyrene in the place of linear polystyrene, which is used in the polystyrene–AlCuCl$_4$ complex solution as described in Section 2.2. The cross-linked polystyrene forms macroretricular (MR)-type resin beads (styrene–divinyl–benzene copolymer, divinylbenzene content 20%; diameter 296–740 μm; average pore diameter 9 nm; surface area 300 m^2/g).

The polystyrene beads (26.1 meq in phenyl groups), copper (I), chloride (22.7 mmol), and aluminum chloride (22.7 mmol) in carbon disulfide (20 mL) are refluxed under dry nitrogen for 6 h. The solvent is thoroughly removed by evaporation, resulting the polstrene–AlCuCl$_4$ complex solid.[12]

The resulting polymer complex solid adsorbs CO rapidly at 25°C, 1 atm, as shown in Fig. 7.3. The equilibrium molar ratio of the adsorbed CO to the admitted copper (I) chloride is 1.07. The amount of adsorbed CO per g of the polymer complex solid is 67 cm^3 (standard temperature and pressure).[13]

After the first adsorption, the polymer complex solid is in contact with

Figure 7.3 Adsorption of CO under 1 atm at 25°C by the polystrene–AlCuCl$_4$ complex solid: the first adsorption (○), the second adsorption (●), and the third adsorption (△); the desorption of absorbed CO is conducted under 7 mm Hg at 25°C for 10 min; prior to the third adsorption, the polymer complex solid is in contact with 10L of nitrogen containing 10 mol % water to the admitted copper (I) chloride.

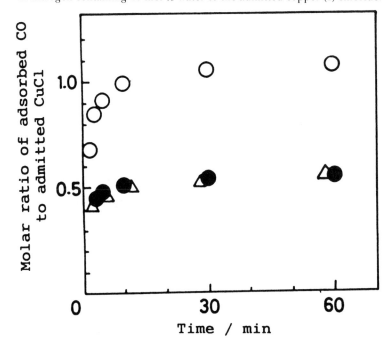

10 L of dry nitrogen at 25°C for 10 min, and then is subjected to a reduced pressure (7 mm Hg) at 25°C for 10 min. In the second adsorption, the polymer complex solid promptly adsorbs CO, and the equilibrium molar ratio of adsorbed CO to the admitted copper (I) chloride is 0.54, as shown by the solid circles in Fig. 7.3. The open triangles in Fig. 7.3 depict the third adsorption of CO. Before the adsorption, the polymer complex solid is contacted with nitrogen containing 10 mol % water to the admitted copper (I) chloride (water content: 5600 ppm) for 10 min, and is kept at 7 mm Hg and 25°C for 10 min. The equilibrium molar ratio of the adsorbed CO to the admitted copper (I) chloride is 0.54, which is identical to the value in the second adsorption. In the subsequent repeated adsorptions, the time courses of CO adsorptions are virtually the same as those for the second and the third adsorptions. Prior to each of these adsorptions, the polymer complex solid is in contact with nitrogen containing 10 mol % water (water content: 5600 ppm).[13]

When the desorption of CO is conducted under 7 mm Hg at 90°C for 30 min, the equilibrium molar ratio of adsorbed CO and that of desorbed CO, respectively, to the admitted copper (I) chloride goes up to 1.07.[12]

The adsorbing capacities of the polymer complex solids prepared with the use of carbon disulfide and dichloromethane remained virtually unchanged after repeated contact of the polymer complex solid with nitrogen gas containing 10 mol % water with respect to the admitted copper (I) chloride. The polymer complex solids obtained with the use of benzene and toluene as preparation solvents, however, exhibited small but gradual decreases in the adsorbing capacities on repeated contact with nitrogen gas containing water.[13]

Scanning electron microscopy (SEM) and energy-dispersive x-ray microanalysis on the cross section of the beads of the polymer complex solids clarify the role of the preparation solvents. In beads of the polymer complex solid prepared by the use of toluene, Al atoms exist largely on or near the surface of the bead, and a number of crystalline deposits are observed in the inner portion of the bead, as shown in Fig. 7.4(a). In the beads of the polymer complex solid obtained with the use of carbon disulfide, Al atoms are distributed uniformly in the cross section of the bead, and no crystalline deposits are seen in the whole cross section, as shown in Fig. 7.4(b). Obviously, benzene and toluene compete with the phenyl groups of the polystyrene and prevent effective bonding of aluminum copper chloride.[13]

4.2. Polystyrene with Amino-Group–CuCl Complex Solid

A polystyrene with amino groups is a macroreticular-type polystyrene resin (Fig. 7.5) having primary and secondary amino groups (resin beads of diameter 0.35–0.55 mm; ion-exchange capacity 2.5 eq L^{-1}; apparent density 650 g L^{-1}).

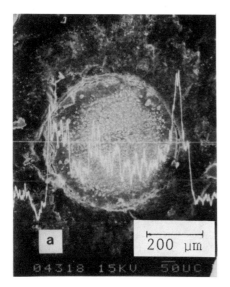 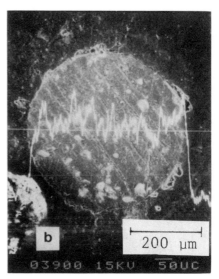

Figure 7.4 SEM photographs and Al atom distribution for the cross section of beads of polystyrene–AlCuCl₄ complex solid prepared by use of (a) toluene and (b) carbon disulfide as solvents.

The polystyrene resin (10.0 g) having amino groups and copper (I) chloride (100 mmol) are stirred in acetonitrile (80 mL) at 20°C for 5 h, and then the liquid phase is removed at 5 mm Hg and 80°C, resulting in the polystyrene with amino group–CuCl complex solid.[14]

The amount of copper (I) chloride bound to the polystyrene resin (10 g) having amino groups increases proportionally with increasing amounts of admitted copper (I) chloride up to 50 mmol. Here almost all of the admitted copper (I) chloride is bound to the resin. Above 50 mmol, the amount of the bound copper (I) chloride approaches a saturation value of 70 mmol. Thus, the amounts of copper (I) chloride bound to the polystyrene resin having amino groups are larger than the ion-exchange capacity (38 mmol) of 10.0 g of the resin.

Figure 7.6 shows the adsorption of CO by the polymer complex solid containing 64 mmol of bound copper (I) chloride.[14] The open circles depict the first adsorption of CO by the polymer complex solid from 9:1 CO/

Figure 7.5 Polystyrene with amino groups in a macroreticular-type resin.

$$-CH_2-CH-CH_2-CH-CH_2-$$

$$-CH-CH_2-$$

$$CH_2NH(CH_2-CH_2NH)_2H$$

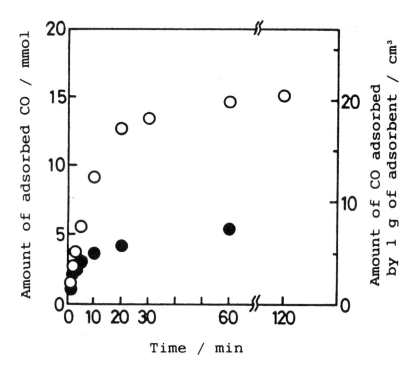

Figure 7.6 Adsorption of CO from 9:1 CO/N_2 mixture under 1 atm at 20°C by polystyrene with amino group–CuCl complex solid containing 64 mmol of CuCl. The desorption of adsorbed CO is conducted under 5 mm Hg at 20°C for 10 min; the first (○) and the second (●) adsorptions.

nitrogen mixture under 1 atm at 20°C. The adsorption is rapid, and the equilibrium amount of adsorbed CO is 15.2 mmol. The adsorbed CO is desorbed by subjecting the polymer complex solid to a reduced pressure (5 mm Hg) at 20°C for 10 min. In the second adsorption (the closed circles), the adsorption is rapid and the equilibrium amount of adsorbed CO is 5.4 mmol. Both the time courses and the equilibrium amounts of adsorbed CO in the third and fourth adsorptions are identical to those in the second adsorption.

Figure 7.7 shows plots of the adsorbing capacity for CO (the open circles) and the adsorbing capacity for carbon dioxide (closed circles) against the amount of copper (I) chloride bound to 10.0 g of the polystyrene resin having amino groups.[14] At 0.0 mmol of copper (I) chloride bound, the resin does not adsorb CO in a measurable amount, and the adsorbing capacity for CO_2 is 8.3 mmol. The adsorbing capacity for CO largely increases with increasing amount of the bound copper (I) chloride, whereas the adsorbing capacity for carbon dioxide monotonously decreases. Consequently, the selectivity of CO adsorption, which is defined by the ratio of the adsorbing capacity for CO to the value for carbon dioxide, significantly increases with

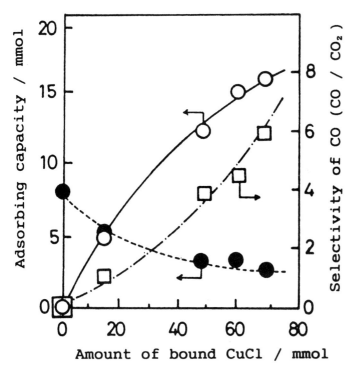

Figure 7.7 Plots of adsorbing capacity (○) for CO, adsorbing capacity (●) for CO_2, and selectivity (□) of CO adsorption against the amount of CuCl bound to the polystyrene resin having amino groups. The selectivity of CO adsorption is defined by the ratio of the capacity for CO to the value for CO_2; the amount of the resin is kept constant at 10.0 g.

an increase in the amount of bound copper (I) chloride, as shown by the open squares. A selectivity of 5.9 is attained at 70 mmol of bound copper (I) chloride.

The polymer complex solid efficiently adsorbs CO by coordination of CO to the copper (I) ions in the complexes between copper (I) chloride and the amino groups of the polystyrene resin. The polystyrene resin without copper (I) chloride adsorbs carbon dioxide both by chemical interaction of the amino groups with carbon dioxide and by physical adsorption of carbon dioxide on the surface of micropores of the resin.

The polymer complex solid contains bound copper (I) chloride in great excess, compared with the ion-exchange capacity of the polystyrene resin, as described previously. Thus, considerable amounts of copper (I) chloride are bound to the resin without formation of complexes with the amino groups of the resin. These copper (I) chloride molecules are connected by chloride bridges with each other and with the copper (I) chloride molecules, which complex with the amino groups of the resin. The layers composed of copper (I) chloride molecules are probably formed on the surface of the micropores of the resin. The x-ray photoelectron spectroscopy on the polymer complex solid is consistent with the formation of layers of copper (I) chloride in the polymer complex solid. The physical adsorption of carbon dioxide to the surface of the resin is suppressed by the layers of copper (I) chloride formed on the surface of the micropores. Chemical adsorption of carbon dioxide due to interactions with the amino groups of the resin is also inhibited by the complex formation between the amino groups and copper (I) chloride.

5. Adsorbents for Ethylene Separation

5.1. Polystyrene–AlCuCl$_4$ Complex Solid

The polystyrene–AlCuCl$_4$ complex solid, which is prepared for CO separation as described in Section 4.1, can also be used for ethylene separation. The polymer complex solid is prepared from polystyrene beads (31.1 meq in phenyl group), copper (I) chloride (26.0 mmol), and aluminum chloride (26.0 mmol) with the use of carbon disulfide (30 mL).[15]

The open circles in Fig. 7.8 depict the adsorption of ethylene under 1 atm at 20°C by the polymer complex solid. The adsorption is rapid, and the equilibrium molar ratio of adsorbed ethylene to the admitted copper (I) chloride is 1.40. The amount of ethylene adsorbed by 1 g of the polymer complex solid is 89 mL (standard temperature and pressure).

Desorption of the adsorbed ethylene is conducted by subjecting the polymer complex solid to a reduced pressure (8 mm Hg) at 20°C for 10 min. In the second adsorption, shown by the closed circles in Fig. 7.8, the equilibrium molar ratio of the adsorbed ethylene to the admitted copper (I) chloride is 0.29. Both the time courses of the adsorptions and the equilibrium molar ratios in the subsequent repeated adsorptions are identical with those in the second adsorption.

The desorption of the ethylene adsorbed by the polymer complex solid is also effectively carried out by heating the polymer complex solid to 90°C under 1 atm for 10 min. The molar ratio of the desorbed ethylene to the admitted copper (I) chloride is 0.47. In the following adsorption, the molar ratio of adsorbed ethylene to the admitted copper (I) chloride is also 0.47. When the desorption of adsorbed ethylene is conducted under 8 mm Hg at 142°C for 10 min, the molar ratio of the desorbed ethylene to the admitted copper (I) chloride reaches 0.87.

The polymer complex solid exhibits no measurable adsorptions of nitrogen, hydrogen, and methane under 1 atm at 20°C.

A 0.47:0.53 ethylene/CO mixture is contacted with the polymer complex solid prepared by using 12.0 mmol of copper (I) chloride under 1 atm at

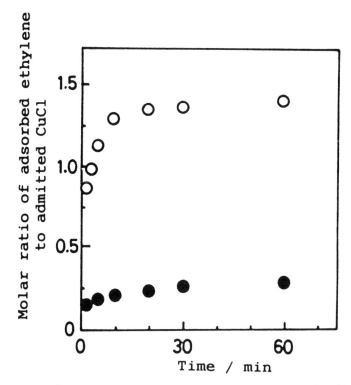

Figure 7.8 Adsorption of ethylene under 1 atm at 20°C by the polystyrene–AlCuCl$_4$ complex solid. The desorption of adsorbed ethylene is conducted under 8 mm Hg at 20°C for 10 min after the first adsorption; the first adsorption (○) and the second adsorption (●).

20°C. The polymer complex solid rapidly adsorbs both ethylene and carbon monoxide from the gas mixture, and the equilibrium amounts of adsorbed ethylene and adsorbed carbon monoxide are 9.0 and 1.8 mmol, respectively.[15] Thus the adsorption selectivity of ethylene to CO attains 5.0.

5.2. Polystyrene–AgAlCl$_4$ Complex Solid

The polystyrene forms macroreticular-type resin beads, as described in Section 4.1. The polystyrene beads (46.5 meq in phenyl group), silver chloride (23.2 mmol), and aluminum chloride (23.2 mmol) in carbon disulfide (20 mL) are magnetically stirred and refluxed for 6 h under dry nitrogen in a 100 ml flask covered with aluminum foil. Then carbon disulfide is thoroughly removed from the mixture by evaporation at 4 mm Hg, 40°C for 2 h, resulting in the polystyrene–AgAlCl$_4$ complex solid.[16]

As shown by the open circles in Fig. 7.9, the adsorption of ethylene (containing 0.6 mol % water) under 1 atm at 20°C by the polymer complex solid is very rapid. The molar ratio of adsorbed ethylene to the admitted silver chloride is 0.65 in 5 min and 1.01 in 120 min. The adsorbed ethylene is released by subjecting the polymer complex solid to a reduced pressure of 8 mm Hg at 20°C for 10 min. On the second contact with ethylene (containing 0.6 mol % water) under 1 atm at 20°C, the polymer complex solid quickly adsorbs ethylene again, and the molar ratio of adsorbed ethylene to the admitted silver chloride is 0.98 in 120 min (closed circles). In the following three adsorption–desorption cycles, the amount of adsorbed ethylene is maintained at the same value as that in the second adsorption. Thus the polymer complex solid repeatedly exhibits adsorption of ethylene without apparent deterioration, even in the presence of a considerable amount of water.

The polymer complex solid shows no measurable adsorption of carbon monoxide under 1 atm at 20°C, as depicted by the closed squares in Fig. 7.9. Thus, the polystyrene–AgAlCl$_4$ complex solid adsorbs ethylene selectively.

Figure 7.9 Adsorption of ethylene under 1 atm at 20°C by the polystyrene–AgAlCl$_4$ complex solid. The desorption of adsorbed ethylene is conducted under 8 mm Hg at 20°C for 10 min; the first adsorption (○), the second adsorption (●), the third (□), and the fourth (▲) adsorption; the adsorption of CO (■) under 1 atm at 20°C.

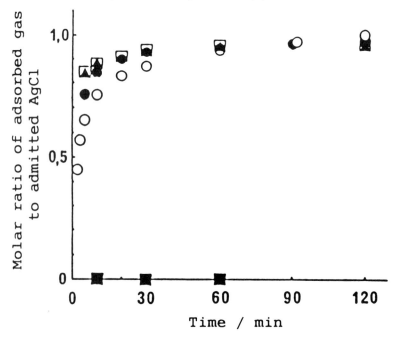

5.3. Polystyrene with Amino-Group–CuCl Complex Solid

The polystyrene with amino-group–CuCl complex solid, which is prepared for CO separation as described in Section 4.2, can also be used for ethylene separation. The macroreticular type of polystyrene resin (10.0 g), having primary and secondary amino groups and copper (I) chloride (152 mmol), are stirred in 1:1 acetonitrile–water mixture (80 mL) at 20°C for 4 h, and then the liquid phase is thoroughly removed at 80°C, 7 mm Hg for 4 h, resulting in polystyrene with the amino-group–CuCl complex solid.[17]

The resulting polymer complex solid adsorbs ethylene rapidly under 1 atm at 20°C, 13.4 mmol in 60 min, and 15.1 mmol in 120 min. Desorption of the adsorbed ethylene is conducted by subjecting the polymer complex solid to a reduced pressure (5 mm Hg) at 80°C for 30 min. The second adsorption is also rapid, and the amount of adsorbed ethylene is 11.8 mmol in 120 min. In the third and fourth adsorptions both the rate of adsorption and the equilibrium amount of ethylene adsorbed are almost identical to the values for the second adsorption.[17]

When the polymer complex solid contacts ethane under 1 atm at 20°C, the adsorption of ethane adsorbed is rapid, but the equilibrium amount of ethane adsorbed is only 1.9 mmol.[17]

Figure 7.10 depicts plots of the amounts of adsorbed ethylene and ethane as a function of the amount of copper (I) chloride bound to the polystyrene resin having amino groups. In these examples the amount of polystyrene resin having primary and secondary amino groups is kept constant at 10 g. The amount of adsorbed ethylene increases with an increasing amount of the bound copper (I) chloride. In contrast, the amount of ethane adsorbed decreases with an increasing amount of bound copper (I) chloride. As a result, the selectivity for ethylene adsorption, defined as the ratio of the amount of adsorbed ethylene to adsorbed ethane, significantly increases as the amount of bound copper (I) chloride increases.[8,17] The selectivity reaches 7.9 when the amount of bound copper (I) chloride is larger than 70 mmol.

The polymer complex solid efficiently adsorbs ethylene by coordination of ethylene to the copper (I) ions in the complexes between copper (I) chloride and the amino groups of the polystyrene resin. The amounts of copper (I) chloride bound to the polystyrene resin having amino groups are larger than the ion-exchange capacity (38 mmol) of 10.0 g of the resin. The excess of the bound copper (I) chloride forms layers of copper (I) chloride on the surface of the micropores of the resin, as described in Section 4.2. Thus, the physical adsorption of ethane on the surface of the resin is suppressed by the layers of copper (I) chloride formed on the surface of the micropores.

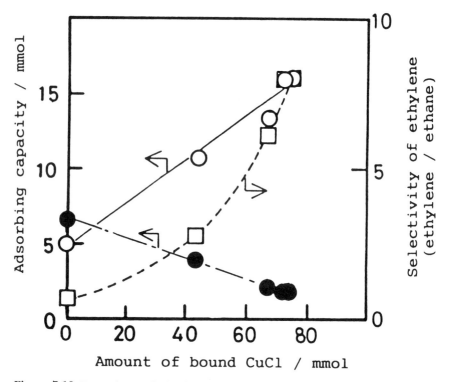

Figure 7.10 Dependence of adsorbing capacity of polystyrene with amino-group–CuCl complex solid on the amount of bound CuCl. Ethylene adsorption (○), ethane adsorption (●), and ethylene selectivity (□); polystyrene resin, 10 g, adsorption under 1 atm at 20°C.

Table 7.1 collects the amount of gas adsorbed both by the polymer complex solid and by the polystyrene resin itself on contact with various gases under 1 atm at 20°C for 2 h. The polymer complex solid composed of copper (I) chloride (70 mmol) and polystyrene resin with amino groups (10 g) adsorbs ethylene, propylene, and carbon monoxide selectively compared with ethane, propane, and carbon dioxide, respectively. The polystyrene resin having amino groups without copper (I) chloride (10 g) adsorbs a large amount of carbon dioxide, and considerable amounts of ethane and propane.

A macroreticular-type polystyrene resin having tertiary amino groups was prepared by reaction of the polystyrene resin having primary and secondary amino groups with aqueous formaldehyde solution in formic acid at 90°C for 16 h.[18]

A polystyrene with tertiary-amino-group–CuCl complex solid has been prepared from polystyrene resin having a tertiary amino group (10 g) and copper (I) chloride (150 mmol) for ethylene separation.[18] The amount of

Table 7.1 Adsorbing Capacity of the Polymer Complex Solid Composed of CuCl (70 mmol) and Polystyrene Resin with Amino Group (10 g)

Gas	Adsorbiing Capacity (mmol)	
	Polymer Complex	Resin without CuCl
Methane	0.1	0.1
Ethylene	15.1	4.9
Ethane	1.9	6.7
Propylene	9.9	8.1
Propane	2.8	7.6
Carbon monoxide	15.8	0.1
Carbon dioxide	2.3	13.5
Hydrogen	0.1	—
Nitrogen	0.1	—
Argon	0.1	—

copper (I) chloride bound to the polystyrene resin having tertiary amino groups (10 g) is 50 mmol.

In the first adsorption, the polymer complex solid rapidly adsorbs ethylene under 1 atm at 20°C and reaches the equilibrium amount of the adsorbed ethylene (5.5 mol) in 30 min. The adsorbed ethylene is desorbed by subjecting the adsorbent to a reduced pressure (3 mm Hg) at 20°C for 30 min. In the second adsorption, the adsorption is also rapid and the equilibrium amount of the adsorbed ethylene is 5.0 mmol. The selectivity of ethylene to ethane is 5.5.

Equilibrium for ethylene adsorption by the present adsorbent is attained in about 30 min. This value is considerably smaller than that for the adsorbent prepared from 150 mmol of copper (I) chloride and 10.0 g of the polystyrene resin having primary and secondary amino groups.

The desorption of the adsorbed ethylene for the present adsorbent is much more efficient than that by the adsorbent prepared by use of the polystyrene resin having primary and secondary amino groups. For the case of the present adsorbent, 91% of the adsorbed ethylene is desorbed by subjecting the adsorbent to a reduced pressure (3 mm Hg) at 20°C for 30 min. In the case of the adsorbent with the polystyrene resin having primary and secondary amino groups, however, the corresponding value is 46%.

REFERENCES

1. A. L. Kohl and F. C. Riesenfeld, *Gas Purification*, 3rd ed., Gulf Publishing Co., Houston (1979).
2. D. J. Haase and D. G. Walker, *Chem. Eng. Prog.* **70**, 74 (1974).

3. D. J. Walker, *Chemtech*, p. 308 (May, 1975).
4. T. Inukai, *Sekiyu Gakkai Shi* **20**, 317 (1977).
5. H. Hirai, S. Hara, and M. Komiyama, *Bull. Chem. Soc. Jpn.* **59**, 109 (1986).
6. N. Toshima, K. Kanaka, M. Komiyama, and H. Hirai, *J. Macromol. Sci.-Chem.* **A25**, 1349 (1988).
7. H. Hirai, K. Kanaka, and M. Komiyama, *Chem. Lett.*, 201 (1986).
8. H. Hirai, *J. Macromol. Sci.-Chem.* **A27**, 1293 (1990).
9. H. Hirai, S. Hara, and M. Komiyama, *Chem. Lett.*, 861 (1984).
10. S. Hara, Ph.D. Thesis, The University of Tokyo, Tokyo (1986).
11. H. Hirai, S. Hara, and M. Komiyama, *Chem. Lett.*, 257 (1986).
12. H. Hirai, S. Hara, and M. Komiyama, *Bull. Chem. Soc. Jpn.* **59**, 1051 (1986).
13. H. Hirai, S. Hara, and M. Komiyama, *Bull. Chem. Soc. Jpn.* **60**, 385 (1987).
14. H. Hirai, K. Wada, K. Kurima, and M. Komiyama, *Bull. Chem. Soc. Jpn.* **59**, 2553 (1986).
15. H. Hirai, M. Nakamura, S. Hara, and M. Komiyama, *Bull. Chem. Soc. Jpn.* **59**, 3655 (1986).
16. H. Hirai, S. Hara, and M. Komiyama, *Angew. Makromol. Chem. Lett.* **130**, 207 (1985).
17. H. Hirai, K. Kurima, K. Wada, and M. Komiyama, *Chem. Lett.*, 1513 (1985).
18. M. Komiyama, K. Kurima, and H. Hirai, *Angew. Makromol. Chem.* **156**, 187 (1988).

Subject Index

A
AFS, 56
Absorbent for CO separation, 222
Absorbent for ethylene separation, 226
Absorption, 148
Acetylene coordination, 201
Acetylene permeability coefficient, 208
Acrylic acid resin, 153
Activated charcoal, 149
Activation energy for diffusion, 128
Activation energy, for gas permeation, 124
Activation energy, O_2 transport, 211
Adsorbent for CO separation, 227
Adsorbent for ethylene separation, 232
Adsorbing capacity, polymer complex, 237
Adsorption, 148
Adsorption curve
  NO by CR-Fe, 161
  NO by dispersion, 172
  NO by mixed-valence, 167
Adsorption isotherm, 164
Adsorption of CO, 152
Adsorption of NO by dispersion, 171
Adsorption rate, 162
  dry and wet, 175
Adsorption scheme of NO by FeOOH, 178
Adsorption selectivity, ethylene/CO, 233
Adsorption, nitrogen monoxide, 160
Advantages of polymers, 10
$AgAlCl_4$, 226
$AgAlCl_4$ complex solid, 233
Aggregation state, 69, 77, 84
Aggregation structure, 52
Aging effect, physical, 133
Air separation, 4
Air, oxidation of Fe(II) by, 170
Air-facing surface, 56
$AlCuCl_4$, 223
$AlCuCl_4$ complex solid, 232
Allosteric binding, 216
Alumina, 149
Aluminum copper(I) chloride, 223
Aluminum silver chloride, 226
Amino-polystyrene, 228

Amino-polystyrene-CuCl complex, 235
Ammonia selectivity, 191
Amphiphile, 62
Anion exchange ability, resin complex, 159
Antioxidant, polyacetylene, 123
Apparent activation energy, 26
Apparent volume, resin complex, 157
Aqueous dispersion vs. dry system, 174
Aqueous dispersion, resin complex, 171
Argon, permeation in PSHD, 79
Arrhenius plot
  PC/EBBA, 91
  PFTA, 101
  PVC/CHOB, 96

B
BPBB, 97
Bilayer, 61
Bilayer membrane, 52
Bimolecular lamellae, 55
Bishistidinecobalt, 87
Bispentylbenzoyloxybenzene, 85
Bissalicylidenecobalt, 187
Boiling point of gas, 127
Bound copper(I) chloride, 229
Butane, 95

C
CHOB, 85, 95
CO separation, absorbent, 222
CO separation, adsorbent, 227
CO/N_2 selectivity, 191
CO/N_2 separation, 224
CO_2, physical adsorption, 231
CPB, 94
CR, 153
CR-Al(III), SEM, 155
CR-Fe complex, uv-vis, 163
Cr-Fe, ESR, 163
CR-Fe(II), 153
  aqueous dispersion, 171
  effective Fe(II), 165
  preparation, 161
  surface area, 155

239

CR-Fe(III), 153
CR-metal complex, porosity, 157
Carbon dioxide, 10, 231
Carbon disulfide, 228
Carbon monoxide, 148, 221
Carbon monoxide transport, 191
Carrier-media transport, 185
Cascade, 35
Catalyst, polyacetylene synthesis, 117
Categories of gas separation, 6
Cellulose acetate membrane, 40
Chain packing, 74, 81
Channel, 93
Charcoal, 149
Charge transfer band, 162
Charge-transfer interaction, 224
Chelate resin, 149
Chelate resin complex, porosity, 157
Chemical adsorption, 149
Chemical dissolution, 196
Cis isomer, 68
Cobalt Schiff-base complex, 187, 206
Cobalt dry-cave complex, 187
Cobaltporphyrin, 204
Cobaltporphyrin complex, 193
Cohesive energy, 154
Complex formation constant, 165
Complex formation constant, temp. dep., 172
Complex formation, polystyrene-metal, 224
Complex membrane, 183
Composite film, 52, 62
Composite membrane, 206
Concentrating effect, 11
Concentration dependence, 19, 26
Concentration dependence, time lag, 112
Concentration polarization, 31
Concurrent flow, 33
Constant cut cascade, 37
Continuous membrane column, 39, 43
Coordinated acetylene, 201
Coordinated dinitrogen, 200
Coordination of NO in water, 175
Coordination of NO to CR-Fe, 164
Copper(I) chloride complex, 228
Copper-ammonium-salt process, 222
Countercurrent flow, 33
Countercurrent effect, 33
Countercurrent permeator, 44
Cross flow, 33
Cryogenic method, 5
Cryogenic separation process, 222
Crystal-nematic transition, 84
$CuAlCl_4$, 151
CuCl complex, 228
CuCl-polymer complex, 235
Cyanoheptyloxybiphenyl, 85
Cyanopentylbiphenyl, 94
Cyclopentadienylcarbonylmanganese, 200

D
DSC, 85
Desorption of NO, 168
Dielectric anisotropy, 94
Diffusion coefficient, 26
  O_2, 211
  PSHD, 80
  integral, 111
  polyacetylene, 127
Diffusivity, 52, 99
Diluting effect, 11
Dinitrogen complex, 200
Dioxocobaltporphyrin, 199
Diphenylpropane, 224
Disalicylidenecobalt, 190
Dispersion system, 173
Double fluorocarbon amphiphile, 63
Dry adsorbent, 160
Dry and wet, adsorption rate, 175
Dry polymer membrane, 210
Dry system vs. aqueous dispersion, 174
Dual-mobility model, 27
Dual-mode sorption, 23, 189
Dual-mode sorption model, 113
Dual-mode transport, 209
Dual-mode transport model, 114
Durability of adsorbent to O_2, 169

E
EBBA, 85
EDTA, 160
ESR, CR-Fe, 163
Effective Fe(II), surface area, 166
Effective amount of Fe(II), 165
Effective diffusion coefficient, 27
Electric field, 94
Electrostatic repulsion, 159
Endon-type coordination, 196
Endon-type dinitrogen, 200
Enrichment of amphiphile, 62
Enthalpy of NO adsorption, 174
Enthalpy of solution, 22
Entropy of NO adsorption, 174
Environment, 10
Equilibrium constant
  complex formation, 225
  dry vs. wet, 175
  NO adsorption, 173
Ethane, adsorption, 235
Ethoxybenzylidenebutylaniline, 85
Ethylene, 221
Ethylene/CO selectivity, 233
Ethylene/ethane selectivity, 191
Ethylenediaminetetraacetate, 160
Exhaust gas, 5

F
Facilitated transport, 185, 192, 203
FeOOH system, 177
Ferric hydroxide, 177

Index

Fick's first law, 17
Fick's law, 78
Fick's second law, 18
Fickian sorption, 89
Field effect, 11
Fixed carrier, 207
Flash photolysis, 203
Flory-Huggins equation, 22
Flow model, 15
Flow scheme, 33
Fluorinated poly(TMSP), 130
Fluorine treatment of poly(TMSP), 130
Fluorocarbon, 85
Fluorocarbon amphiphile, 52, 55
Fluorocarbon bilayer, 62
Fluorocarbon monomer, 100
Fractional free model, 115
Fracture surface, 86
Free volume, 84, 92, 205
Free volume model, 115
Free-molecule flow, 15
Free-standing film, 121
Frozen free volume, 205

G

Gas diffusion, 24
Gas permeability, polyacetylene, 124
Gas permeation, theory, 15
Gas solution, 21
Gas sorption by metal complex, 150
Gas-permeation property, 86
Gas-reactive molten salt, 191
Gas-separation scheme, 28
Gel-type bead, 151
Glass transition concentration, 23
Glass transition temperature, 20
   polyacetylene, 121
Glassy polymer, 22, 26, 83
Glassy state, 66, 154
Graft content, poly(TMSP), 131
Granular structure, 75

H

Head separation factor, 32
Heat of solution, 22
   polyacetylene, 128
Helium, permeation in PSHD, 79
Hemoglobin, 183
Henry mode, 209
Henry's constant, 174
Henry's law, 22
Heterogeneity, 74
Heterogeneous aggregation, 77
Heterogeneous matrix, 81
Hexane, 98
High molecular weight, 118
Hollow fiber, 151
Homogeneous aggregation, 77
Hopping pathway, 211

Hydrocarbon gas, sorption in membrane, 89
Hydrocarbon isomer, 92
Hydrogen, permeation in PSHD, 79

I

IDA, 153, 160
Ideal cascade, 37
Ideal separation factor, 113
Iminodiacetic acid moiety, 153
Immobilization of metal complex, 160
Immobilization of metal ion, 150
Induction period, 204
Inorganic adsorbent for NO, 177
Integral diffusion coefficient, 111
Intermolecular channel, 93
Intermolecular distance, 98
Intermolecular interaction, 97
Ion etching, 74
Ion-exchange capacity, aminopolystyrene, 228
Iron porphyrin complex, 184, 206
Irreversible oxidation, 199, 206
Isomer
   cis-trans, 68
   hydrocarbon, 92

J

Jarosite, 177

K

Kinetic limitation, 214
Knudsen flow, 15
Kr monitoring system, 42

L

LB film, 52
LC, 84
Langmuir adsorption, 23
Langmuir equation, 162
Langmuir isotherm, 196, 201
Langmuir mode, 209
Langmuir plot, 165
   mixed-valence complex, 168
Langmuir-Blodgett film, 52
Laser flash spectroscopy, 196
Layer structure of jarosite, 178
Lifetime for oxygen-binding, 198
Ligand-exchange reaction, 175
Liquid crystal, 84
Liquid membrane, 185, 187, 192
Liquid-crystalline phase, 206
Liquid-crystalline state, 62
Local diffusion coefficient, 18
Low-spin complex, 162

M

MWD, polyTMSP, 118
Macroporous polystyrene, 152
Macroreticular-type resin bead, 227

Manganese complex, 200, 208
Matrix model, 24
Mean diffusion, 19
Mean permeability coefficient, 18
Mechanical property, polyacetylene, 121
Membrane method, 7
Membrane permeation system, 46
Membrane permeation, theory, 15
Metal complex for gas sorption, 150
Metal complex solution, 222
Metal ion, effect on porosity, 156
Metalloporphyrin complex, 198
Microgel, 151
Microgravimetric measurement, 193
Micropore of resin, 231
Microvoid, 20
Miscibility, 156
Mixed multibilayer, 62
Mixed-valence complex, 166
Mobility selectivity, 113
Molecular aggregation, 66
Molecular diameter, 128
Molecular filtration, 92
Molecular orientation, 92
Molecular orientation effect, 84
Molecular packing, 78
Molecular weight distribution, polyTMSP, 118
Multibilayer, 62
Multicomponent bilayer, 63
Mutual diffusion coefficient, 110
Myoglobin, 183

N
NMR, poly(TMSP), 138
NO
  adsorption, 160
  adsorption by dispersion, 171
  adsorption by mixed-valence complex, 167
  coordination to CR-Fe, 164
  desorption, 168
  inorganic adsorbent, 177
  simultaneous removal, 176
NO adsorption
  enthalpy, 174
  entropy, 174
NTA, 160
Naylor-Backer, 34
Need for gas separation, 3
Nematic-isotropic transition, 84
Nitrilotriacetate, 160
Nitrogen, 4
Nitrogen complex, 200
Nitrogen monoxide, 153
  adsorption, 160
  adsorption, dispersion, 171
  simultaneous adsorption, 176
Nitrogen permeability coefficient, 203, 208
Nitrogen permeation in PSHD, 79
Nitrogen transport, 207
Nitrogen-binding rate constant, 201
No-mixing cascade, 37
Nonporous membrane, 17
Number of bilayer, 61

O
Occupied volume, 72
Optimum oxygen-binding affinity, 189
Orientation, 92
Outermost layer, 61
Oxidation, 199
Oxidation of Fe(II), 169
Oxidation of cobaltporphyrin, 206
Oxidation of resin complex, 170
  mechanism, 171
Oxidation, polyacetylene, 123
Oxy-deoxy cycle, 198
Oxygen, 4
Oxygen carrier, 186
Oxygen complex, 199, 207
Oxygen diffusion coefficient, 211
Oxygen enrichment, 100
Oxygen permeability coefficient, 125, 203
Oxygen permeation in PSHD, 79
Oxygen permselectivity, 101
Oxygen sorption, 195
Oxygen transport, 186
Oxygen-binding, 184
Oxygen-binding affinity, 189
Oxygen-binding rate constant, 198
Oxygen-coordinated complex, 207
Oxygen-enrichment membrane, 102
Oxygen/nitrogen, 62
Oxygen/nitrogen selectivity, 186, 207

P
PAS, 8
PFTA, 100
PSHD, 66, 69, 78
PVC, 85
Packing coefficient, 73
Parallel-type separation cell, 41
Partial immobilization model, 27
Patchy overlayer model, 58
Pentane, 97
Perfect mixed flow, 33
Perfluorotributylamine, 85, 100
Permeability coefficient, 9, 19, 111
  Co complex, 203
  Mn complex, 208
  O_2/N_2, 64
  polyacetylene, 125
  polymers, 83
  PSHD, 80
  PVC/CPB, 95
Permeability coefficient ratio, 95, 100
Permeability constant, 111

Permeability of PSHD, 79
Permeability theory, 29
Permeated stream, 34
Permeation curve, 109
Permeation of gas in poly(TMSP), 132
Permeation of gas mixture in poly(TMSP), 136
Permeation rate, 111
Permeation theory, 15
Permselective membrane, 83
Permselectivity
  hydrocarbon isomer, 93
  O_2/N_2, Co porphyrin, 215
  oxygen, 101
  oxygen/nitrogen, 62
  poly(TMSP), 136
  polyacetylene, 124
Phase transition, 84, 90
Photodissociation of oxygen, 197
Physical adsorption, 149
Physical adsorption of CO_2, 231
Physical adsorption of ethane, 235
Physical pore, 151
Physical sorption, 196
Pi coordination, 225
Plasma polymerization, poly(TMSP), 130
Poiseuille flow, 15
Polarizing optical micrograph, 87
Poly(acetylene), 117
  substituted, 119
  synthesis, 120
Poly(dimethylsiloxane), 129
Poly(oxydimethylsilylene), 83
Poly(spiroheptadiene), 66
Poly(styrene), adsorbent, 152
Poly(styrene) resin, porous, 149
Poly(TMSP), 117
  analogue, 119
  modification, 130
  NMR, 138
  permeability, 132
  structure, 135
Poly(trimethylsilylpropyne), 132
Poly(vinyl alcohol), 53
Poly(vinyl chloride), 85
Polyamine carboxylate, 160
Polymer complex membrane, 192
Polymer complex solid adsorbent, 237
Polymer effect, 11
Polymer manganese complex, 208
Polymer membrane, 9
Polymer property, polyacetylene, 121
Polymer synthesis, polyacetylene, 117
Polymer-CuCl complex, 228
Polymer-metal complex, 148, 183
Polymer-metalloporphyrin complex, 193
Polymer/LC composite film, 84
Polymer/LC membrane, 90
Polymer/molecular membrane, 52
Polymer/multibilayer membrane, 62

Polystyrene with amino-group, 228, 235
Polystyrene-AgAlCl$_4$ complex, 226, 233
Polystyrene-AlCuCl$_4$ complex, 223, 225, 232
Polytrimethylsilylpropyne, 205
Pore size distribution, CR-Fe, 158
Porosity, 150
  resin complex, 156
  solvent treatment, 153
Porous membrane, 15
Porous polymer, 150
  SEM, 152
Porous resin, 149
Porous resin complex, adsorption of NO, 160
Porphyrin complex, 184, 193
Potential barrier, 68
Potential energy calculation, 66
Precipitant, 151
Present-deBethune effect, 16
Pressure ratio, 30
Pressure swing adsorption, 8
Pressure swing method, 222
Prism, 4, 8
Process design, 28
Propane, adsorption, 236
Property, polyacetylene, 121
Propylene, adsorption, 237

R
Random-coil model, 77
Recombination of oxygen, 197
Recovery of resin, 176
Recyclic use of adsorbent, 168
Recycling scheme, 39
Relaxation magnitude, 71
Repeated use of adsorbent, 169
Reprecipitation, 154
Resin complex
  anion exchange, 159
  oxidation, 170
  porosity, 156
  preparation, 160
  structure, 159
  surface area, 160
Restricted diffusion, 17
Reversible oxygen binding, 193
Rigid polymer, 66
Rigid sequence, 66
Rubberlike state, 154
Rubbery polymer, 21, 25, 83

S
SEM, polystyrene-AlCuCl$_4$ complex, 229
SEM, porous polymer, 152
SO_2 adsorption, CR-Fe(II), 176
SO_2, simultaneous removal, 176
Scanning electron micrograph, 56, 86
Scanning electron microscope, 54
Selective sorption, 192

Selectivity, 112
  CO adsorption, 230
  ethylene/CO, 233
  ethylene/ethane, 235
  mobility, 113
  solubility, 113
Semirigid sequence, 66
Separation cascade, 35
Separation efficiency, 31
Separation factor, 29, 32, 64
Separation ratio, 94
Separation stage, 30, 32
Sequence model, 77
Series-type separation cell, 41
Sideon-type coordination, 201
Silica gel, 149
Silicone membrane, 205
Silicone rubber, 40
Simultaneous adsorption, 176
Simultaneous removal, 176
Slip flow, 16
Solid membrane, 192, 203
Solid-state NMR, poly(TMSP), 138
Solubility, 52
  nitrogen in PFTA, 101
  of polyacetylene in solvent, 123
  oxygen in PFTA, 101
Solubility coefficient, 18, 19
  PSHD, 81
  polyacetylene, 127
Solubility mechanism, 91
Solubility parameter, solvent, 154
Solubility selectivity, 113
Solution-diffusion mechanism, 109
Solvent effect, 165
  porosity, 154
Solvent treatment, porosity, 153
Solvent-free membrane, 192
Sorption isotherm, 21, 89
  poly(TMSP), 135
Sorption method, 8
Sorption model, 113
Sorption of gas by metal complex, 150
Sorption of gas in poly(TMSP), 134
Sorption-desorption curve, 89
Specific surface area
  effective Fe(II), 166
  resin complex, 157
  solvent effect, 155
Specific volume, 20
Stage separation factor, 32, 41
Steady-state permeation rate, 111
Steric effect, 11
Stress-strain curve, 88
Styrene sulfonate resin, 153
Styrene-divinylbenzene copolymer, 150
Substituent effect, 124
Substituted polyacetylene, 117
  permeability, 124

mol. weight, 118
property, 121
solubility, 123
synthesis, 117
Sulfur dioxide, simultaneous adsorption, 176
Surface area, 150, 162
Surface area of resin complex, 160
Surface chemical composition, 56, 62
Surface coverage, 59
Surface diffusion, 216
Surface flow, 17
Surface molecular aggregation, 60
Surface structure, 52
Symmetric separation cascade, 37
Synthesis
  polyacetylene, 117
  porous polymer, 151

T
TMSP, 117
Tail separation factor, 32
Takeoff angle, 57
Temperature dependence, NO adsorption, 172
Ternary composite, 100
Thermal fluctuation, 99
Thermal-molecular motion, 72
Thermal stability, polyacetylene, 122
Thermal swing method, 222
Thermodynamic parameter
  N_2, 202
  O_2 binding, 196
  acetylene, 202
Thermogravimetric analysis, polyacetylene, 122
Time dependence, 133
Time lag, 111
Total immobilization model, 26
Total pore volume, CR-Fe, 158
Trans isomer, 68
Transition temperature, 98
Transition-metal catalyst, 117
Transport mechanism, 209
Transport model, 114
Transport phenomenon, 15
Trimethylsilylpropyne, 117
True separation factor, 30
Two-membrane scheme, 38
Two-way interaction, 225

U
Unrelaxed volume, 20

V
van't Hoff plot, 174
Viscous flow, 15

W

WAXD, 97
WK resin, 159
Washing solvent, 154
  CR-Fe(II), 165
Wide-angle diffraction pattern, 55

X

XPS, 52
X-ray diffraction, 94
  PVC/CHOB, 96
X-ray diffraction pattern, 54
X-ray photoelectron spectroscopy, 52